Building Decentralized Trust

Victoria L. Lemieux • Chen Feng
Editors

Building Decentralized Trust

Multidisciplinary Perspectives on the Design of Blockchains and Distributed Ledgers

Springer

Editors
Victoria L. Lemieux
School of Information
University of British Columbia
Vancouver, BC, Canada

Chen Feng
School of Engineering
University of British Columbia Okanagan
Kelowna, BC, Canada

ISBN 978-3-030-54413-3 ISBN 978-3-030-54414-0 (eBook)
https://doi.org/10.1007/978-3-030-54414-0

This Springer imprint is published by the registered company Springer Nature Switzerland AG.
The registered company address is: Gewerbestrasse 11, 6330 Cham, Switzerland

Preface

This manuscript is one of the latest examples of how the blockchain and distributed ledger technology (DLT) research and education cluster at the University of British Columbia, "Blockchain@UBC," has been working as a multidisciplinary group of scholars and educators over the past few years while serving as a catalyst for research, teaching, and community building. Blockchain@UBC currently comprises the largest multidisciplinary research and education cluster in Canada and has thus been able to be at the forefront of blockchain and DLT theoretical thinking and development from a multidisciplinary perspective, a perspective that is, arguably, required to avoid the pitfalls of siloed thinking that can lead to unintentional negative consequences sometimes associated with the introduction of emerging technologies.

The opportunity for the present academic collaboration arose when the Blockchain@UBC cluster received support from the Peter Wall Institute for Advanced Studies, also part of the University of British Columbia, through its International Research Roundtable Program. The International Research Roundtable Program, according to Peter Wall Institute, is intended to "foster excellence in research, and serv[e] as a catalyst for collaborative research between international scholars and UBC scholars."[1] The program aims at providing a platform for scholars, community leaders, artists, policy makers, and diverse networks of stakeholders. The roundtable program allows for exploration of a multidisciplinary topic, creating the foundation for innovative research and prompting relevant discussions and advances in science and society. Within the context of this program, Blockchain@UBC leaders proposed to create a 3-day intensive collaborative experience during the summer of 2019, which was the impetus for this volume. We

[1] https://pwias.ubc.ca/program/virtual-roundtables

sincerely hope the volume meets its aim of contributing to innovative research and prompting multidisciplinary discussions and advances in the design of blockchain and distributed ledger systems.

Vancouver, Canada Victoria L. Lemieux
Kelowna, Canada Chen Feng
April 2020

Acknowledgments

VLL and CF would like to thank the Peter Wall Institute for Advanced Studies for providing funding and logistical support for the International Research Roundtable with which this volume began. Without the generous support to multidisciplinary research efforts such as those at Blockchain@UBC, the insights presented in this volume would not have been possible. In particular, we would like to express our appreciation to Bernadette Mah, Program Manager, at the Peter Wall Institute for Advanced Studies.

We would also like to thank Michelle Ho, Blockchain@UBC's Program Coordinator, for her tireless organizational efforts and logistical support.

For the most part, the roundtable participants served as peer reviewers of one another's chapters, following the single blind peer review process we set out during the roundtable; however, in some cases, we felt it necessary to seek out the special expertise of external reviewers. We would like to thank Dr. Darcy Allen, Research Fellow at the RMIT Blockchain Innovation Hub, Melbourne; Dr. Ning Nan, Associate Professor at the Sauder School of Business, University of British Columbia; and Dr. Konstantin Beznosov, Professor of Electrical and Computer Engineering at the University of British Columbia, for ably reviewing draft chapters.

We would also like to express our appreciation to Dr. Marcelo Bravo, who brought his creative vision to the roundtable and the process of producing this volume—without his expert facilitation we would not have had such a successful roundtable nor would we have had nearly as much fun. To our graduate students and postdoctoral research fellows, who served as roundtable facilitators and chapter production coordinators—Danielle Batista, Amir Fard Bahreini, Darra Hofman, Chang Lu, Chris Rowell, and Artemij Voskobojnikov—please accept our special thanks for your hard work and enthusiasm for this project. You always went above and beyond in your support of this process.

We would be remiss if we were not to thank all of the roundtable participants for sharing ideas and knowledge to contribute to this volume, and for their willingness to engage in our experimental strategic design approach.

Finally, we would like to express our appreciation to Jennette Chalcraft as well as the staff at Springer for their excellent editorial assistance.

Contents

Chapter 1
Introduction: Theorizing from Multidisciplinary Perspectives on the Design of Blockchain and Distributed Ledger Systems (Part I)

Victoria L. Lemieux and Marcelo Bravo

1.1 A Comprehensive Perspective: The Opportunity and Need for Blockchain and Distributed Ledger Technology Systems Integrated Theoretical Advancements

Blockchain and distributed ledger technology (DLT) systems, which Michael Casey and Paul Vigna (2018) have called "The Truth Machine", have emerged as a solution to the problem of trust which, at the moment, is experiencing *une crise* (Edelman 2017). Trust is at an all-time low in connection with data and records (as an example, we are bombarded daily with misinformation); in social, political and economic institutions; and in technical systems that are designed for the manipulation of data but not for its protection. Many people no longer trust our institutions, our information systems, nor the information they contain. Moreover, some individuals increasingly mistrust centralized authorities in any form.

Cheney et al. (2009) observe: "*historically, databases…were trusted because they were under centralized control: it was assumed that trustworthy and knowledgeable people were responsible for the integrity of the data.*" But times have changed. As Collomosse et al. (2018) note, trust in archival institutions, traditionally seen as places that could be trusted to preserve the long-term integrity and authenticity of records, has eroded. This erosion of trust is because, in many contexts, those in control of centralized systems have proven to be untrustworthy, manipulating the

V. L. Lemieux (✉)
School of Information, University of British Columbia, Vancouver, BC, Canada
e-mail: v.lemieux@ubc.ca

M. Bravo
School of Public Policy and Global Affairs, University of British Columbia, Vancouver, BC, Canada
e-mail: marcelo.bravo@ubc.ca

V. L. Lemieux, C. Feng (eds.), *Building Decentralized Trust*,
https://doi.org/10.1007/978-3-030-54414-0_1

records and information which they were supposed to be protecting. Nor has decentralization been the answer up until now: data originating from the web or social media have proven to be quite untrustworthy in many cases. We have lacked, as of today, a comprehensive solution to these problems.

Now, however, blockchain and distributed ledger technology (DLT) systems are being advanced as a solution to the global crisis of trust due to their potential as a digital trust infrastructure. Blockchain is characterized by confirmed and validated sets of transactions that are chained together to make tampering difficult and render records immutable. Blockchain and DLT's design and unique capabilities, some argue, circumvent the need for trust, which is why they are sometimes called "trustless" technologies (Kasireddy 2018). In practice, however, blockchain and DLT technology really does not obviate the need for trust. Instead, they offer a new way to substitute what we once relied upon as the basis of trust but which is now viewed as untrustworthy, inefficient or flawed (e.g., long-term social ties, traditional legal contracts, or information supplied by intermediaries) with other sources of trust (e.g., computation). The unique potential of blockchain and DLT technology *vis-a-vis* the problem of trust establishes them as emerging technologies with socio-economic, data records, and technical implications that far exceed most other emerging technologies.

Yet, despite their potential, blockchain and DLT systems are still under theorized and not well understood. We do not fully comprehend, for example, the ways that different aspects of a blockchain or DLT solution interacts with, or creates trade-offs that have an effect on, trust, whether among human social actors or technical system components. This is a gap that this volume seeks to help fill. The following sections of this introductory chapter therefore will explain (1) the opportunity and need for blockchain and DLT system theoretical advancement, (2) the methodological foundation of the collaborative theory building using a design-led approach known as the Strategic Design Method, (3) the interdisciplinary philosophical underpinnings and preparations for crafting the roundtable experience that has given rise to this volume, (4) the experience of the multidisciplinary work that took place during the Peter Wall Institute for Advanced Studies' International Research Roundtable on blockchain and distributed ledger technologies (hereafter referred to as the PWIAS RT), and the subsequent writing and review activities that materialized into this book, and (5) an applied reflection, lessons learned from the process, and future applications.

1.2 Blockchain and DLT Systems' Interacting Trust Layers

This volume began with the theoretical proposition that the design of blockchains and DLT systems as decentralized trust infrastructures can be said to rely upon three interacting "trust layers" (Lemieux et al. 2019): a *social layer*, the layer at which social actors of all types interact with one another and determine how much information they need, and in what form (e.g., by social convention, how much from the blockchain system and how much from other sources external to the

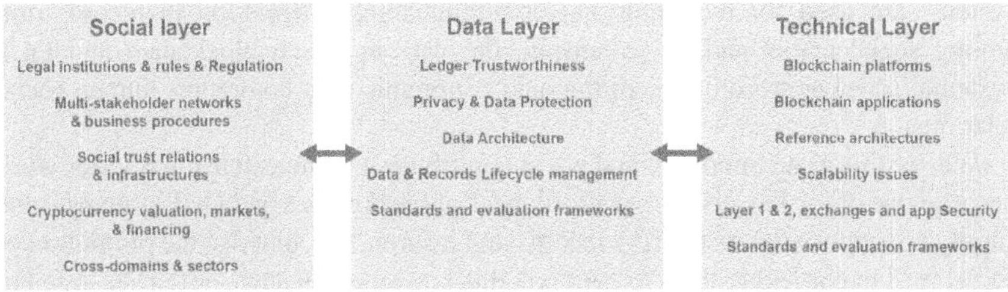

Fig. 1.1 Three-layer trust model of blockchain technology (Lemieux et al. 2019)

system) in order to be able to trust and take action on the basis of that trust; a *data/records layer* that supplies trustworthy (and trusted) information that social actors have decided they need to obtain from the blockchain system to give them confidence to act; and a *technical layer*, being the technical means (e.g., applications, networks, consensus mechanisms) by which social actors interact and create, store and obtain information about those interactions as tamper-resistant and non-repudiable proof of facts about acts (see Fig. 1.1).

Each of these layers interoperates with the goal of achieving trusted transactions. Due to their capacity to alter existing technical, data/records, and social trust relations, blockchain and DLT systems hold the potential to disrupt a myriad of social, political, and economic domains.

Much of the blockchain research to date naturally focuses on the technical aspects of these emerging technologies. In addition, there is a growing body of research focusing on the social, economic, and political potential for transformation of blockchain and DLTs and on the question of blockchains and data privacy. A much more limited amount of research has also been done on the recordkeeping aspects of blockchains. Very little work has been done at the *interstices* of these aspects and their implications for the design of blockchain and DLT systems, though Kannengiesser, Lins, Dehling and Suyaev's work (2019) stands as a notable exception. In that work, the authors observe that blockchain/DLT "*is available in different designs that exhibit diverse characteristics. Moreover, DLT designs have complementary and conflicting characteristics. Hence, there will never be an ideal DLT design for all DLT use cases; instead, DLT implementations need to be configured to contextual requirements*" (p. 1). With this exception, discussions of trade-offs in the blockchain/DLT literature are largely conceptualized in terms of the tension that exists between speed and security (see, e.g., Kiayias and Panagiotakos 2015). Kannegeisser et al. (2019) expand this to considerations of trade-offs among six properties: security, performance, usability, development flexibility, level of anonymity, and institutionalization, drawn from a systematic analysis of extant literature on blockchain and DLT. Although their framework for analysis of the trade-offs among key blockchain and DLT characteristics conceives of blockchain and DLT as socio-technical in nature (for example, they discuss institutionalization and the contextual requirements of different use cases), our work seeks to draw attention to, and raise to a *first order focus of analysis*, the way in which blockchain and DLT systems are used for recording and communicating *information* in record form among social actors and, by extension, the way in which blockchain and DLT systems' novel approach to recording and communication configures human social relations.

We see this as an important and novel contribution to the extension of prior work because one of the key disruptive affordances of blockchains and DLT systems is the supply of supposedly trustworthy records and information aimed at the promotion of social trust in particular; thus, to overlook this key aspect of such systems is to fail to understand their fundamental nature as systems of recordkeeping and, to a large extent, miss the point of blockchain and DLT technology. The interactions among these three aspects (the social; data/records; and technical), we argue, have yet to be understood and explored, though current research certainly points to and confirms the importance of understanding interrelationships in blockchain and DLT system design. By bringing researchers from diverse backgrounds together, the PWIAS RT that gave rise to this volume sought to further unpack relationships by exploring interdependencies among the social, data/records and technological aspects of blockchain and DLT systems which are, at present, not fully appreciated and understood.

Too many novel technologies have been introduced without thought to the way that the technical affordances of the technology impact upon social behavior (think about social media as an example), the way that the business models of new technologies affect user privacy (think about large-scale platforms that gather up our data), or the way in which more efficient machine processing of information affects society (think about AI as an example here). Thus, a core premise of this volume is that a failure to understand and consider the relationships and interdependencies among the interstices of these three aspects—the social, data/records, and technical—of blockchain and DLT systems and, indeed, any emerging technology would likely end in unintended consequences and regret. Through focusing upon and theorizing about the three layers and the interstices and interrelationships among them, the aim of the contributors to this volume is to strengthen the design of blockchain/DLT solutions so that, ultimately, their application may achieve a net beneficial effect on society—as many proponents of these technologies envision blockchain and DLT can do—or, at the very least, avoid some disastrous unintentional introduction of risks to humanity and the environment.

1.3 The Design Approach: Introducing Strategic Design as a Guiding Collaborative Framework

Strategic design has been an ongoing theoretical development in the world of design-led methodologies that has been serving industry in the last two decades, usually referred to as design thinking (Liedtka et al. 2014; Brown 2009; Martin and Christensen 2013). However, its influence has been progressively adopted and adapted in academia and the public sector based on its creative planning possibilities, human-centred approaches, and fast-paced learning applications.

Strategic design, a term explored and expanded in Canada by Professor Moura Quayle—system designer formerly at the University of British Columbia's (UBC) Sauder School of Business and currently at the UBC School of Public Policy and Global Affairs—has served in the last decade to shape and re-think numerous organizational developments and academic initiatives at UBC, in Canada, and abroad (UBC d.studio 2015; UBC Policy Studio 2017). Strategic design advances have been led initially by Sauder's d.studio and more recently by the Policy Studio, both at the University of British Columbia. These units have successfully engaged participants from various academic disciplines, as well as industry, government and the general public. Strategic design can be understood as a design-led method that is human centred, highly collaborative, and that offers an integrative thinking approach that incorporates both critical thinking as well as creative thinking cognitive domains (Quayle 2017).

Strategic design has been used to tackle challenges that span varied academic disciplines, non-academic space, as well as cross-disciplinary endeavors that require a higher level of collaboration, effective reflection, and novel thinking, all of which make it a perfect approach for addressing theoretical gaps in an emerging technology that has an inherently multidisciplinary character.[1] The following is a more in-depth exploration of the particular features of the Strategic Design Method and the rationale for its application to collaborative efforts that supports blockchain and DLT system design theoretical advances.

Strategic design emphasizes a practical, reflective, and a co-generative approach. It starts with an understanding of the interests, assumptions, and guiding values of participants. In the context of a strategic design work, participants act as co-designers of ideas, prototypes, and new initiatives.

Strategic design is an integration of thinking as well as doing; therefore, Quayle, one of its main proponents, departs from the term design *thinking* to assert: "*Design is active. It's a verb. Design is not just about thinking, but about constantly trying and doing*" (Quayle 2017, pp. 75–76). Strategic design works as a learning and collaborative platform for discovery. It allows testing and expanding ideas into new findings, propositions, or models.

The method, according to Bravo's (2019) empirical findings, is in essence an interdisciplinary-inspired methodological approach. Strategic design, he states, "*has benefited from a vast array of disciplinary areas that contribute with core knowledge bringing perspectives that supports problem finding and solution finding*" (p. 94). For instance, strategic design incorporates frameworks and tools from education, psychology, philosophy, management, engineering, and design led disciplines.

[1]Examples include the UBC d.studio "Design Challenge" where multidisciplinary groups of undergraduate and graduate students co-developed multi-sectoral approaches in climate change efforts, or the UBC Policy Studio "Resilient Cities Policy Challenge" where graduate students explored policy programs to strengthen resiliency at the societal level with the City of Vancouver.

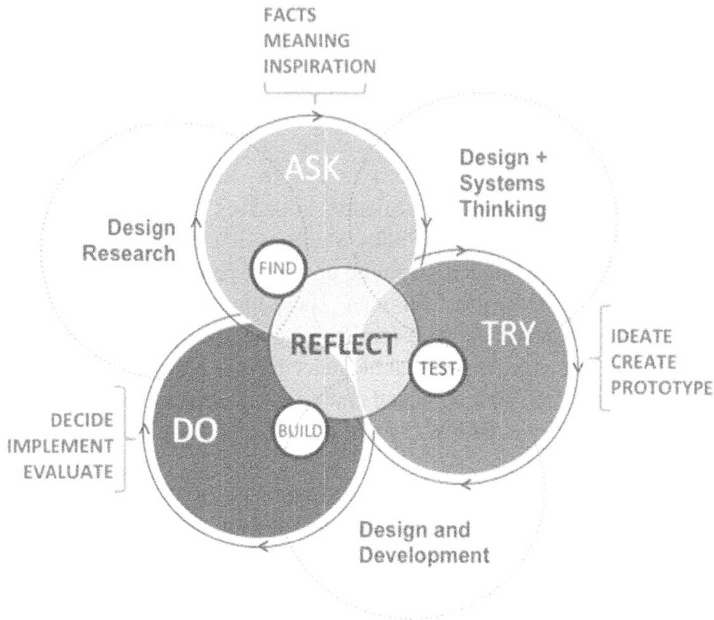

Fig. 1.2 The strategic design process by Quayle and Beausoleil (Reproduced from Quayle 2017, p. 75)

Strategic design is also integrative in ways that deliberately create space for divergent and convergent thinking styles aimed at the exploration of a proposed task or challenge. The method allows space for participants that exhibit different learning styles to be comfortable and participate in co-generative ways through the use and application of toolkits that support learning and engagement. On this, Beausoleil (2016) explored in detail thinking styles and pedagogical approaches that support design-led education that used the Strategic Design Method in learning contexts that targeted enhancement of innovativeness competencies.

Importantly, one of the most relevant features of the method relies on the inclusion of dedicated time for reflection that is deliberately applied throughout design exercises (see Fig. 1.2). Reflection can be programmed as a personal or collective task. As Beausoleil's work revealed, the method can explicitly recognize diverse types of thinking modes and habits, therefore allowing opportunities for participants with different styles to think, reflect, and engage.

Another key feature of the Strategic Design Method is the visual-oriented application of tools and techniques. Facilitators usually propose activities that are highly visual and engaging based on educational research, (see for example Sniukas et al. (2016), Brand (2017), and Galsworth (2018)) supporting the benefits of visual learning capabilities that participants can gradually incorporate in collaborative initiatives.

A final characteristic of strategic design, which appeals to both academia and industry, relates to its highly collaborative nature and how this supports innovation processes. The method is well-adapted to work to address complex problems—usually referred to as "opportunities"—that traditionally exceed well-established academic or domain boundaries (Kolko 2012; Julier and Kimbell 2016; Dunne 2018). As a result, some authors use strategic design to tackle systemic or resilient challenges that involve multidisciplinary approaches and, in some cases, multi-sectoral participation through a series of interactive sessions. Significantly, strategic design or design thinking is a method that several authors propose as a way to achieve "innovations", which Liedtka et al. (2014) define as an "*idea or an invention that is implemented and creates value*" (p. 3).

The list of authors that have used either design thinking or strategic design is extensive and growing. For example, in the domain of business we might find the works of Brown (2009), Martin (2009), Kimbell (2014), Beausoleil (2018), and Liedtka et al. (2014). These and other scholars of strategic design have been testing and applying the use of design-led approaches with projects in government, education and not for profit sectors. Examples can be found in the work of Liedtka (2017), Beausoleil (2016), Brown and Wyatt (2010), Quayle (2017), Bellefontaine (2013), Kolko (2012), Dunne (2018), and studios and consultancy agencies such as IDEO, Helsinki Design Lab, Dk Mind Lab City, the UK's Design Council, City of Vancouver Solutions Lab, amongst others.

The following section now turns to discussing the application of strategic design as a methodological approach for the PWIAS RT, held in Vancouver in 2019 to explore and theorize the nature and interrelationships of the proposed three layers of blockchain and DLT systems.

1.4 Interdisciplinary and Multidisciplinary Underpinnings, and the Design of the Collaborative Experience

As introduced at the beginning of this chapter, the opportunity to include a design-led methodology to foster a unique learning and collaborative theory-building experience was the product of the collaboration of Blockchain@UBC and the Peter Wall Institute through its International Research Roundtable program. A specific theme was defined: "The Truth Machine: Exploring the Social, Records and Technical Potential and Pitfalls of Blockchain and Distributed Ledger Technologies". The academic leaders of the Blockchain@UBC, a University of British Columbia blockchain education and research cluster, carefully selected and invited global thought leaders in blockchain and DLT systems. A key undertaking was to be able to form a broad-multidisciplinary group in terms of academic backgrounds, university affiliations, country of origin, and topic of research in blockchain and DLT, with the interest to share and co-generate knowledge in an intensive, fast-paced collaborative process.

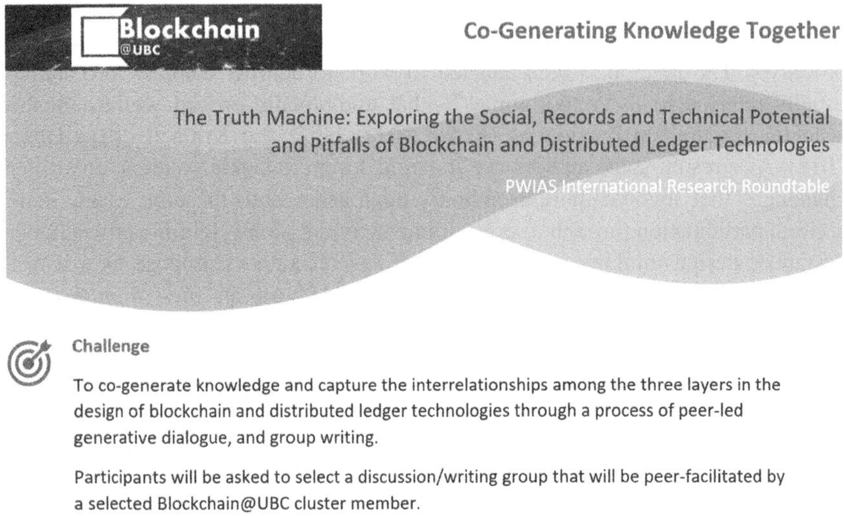

Fig. 1.3 Preparatory activities—meeting with teamwork facilitators

Once the invitation was sent and responses received, 28 individuals had confirmed their participation and willingness to travel to Vancouver, BC for the PWIAS RT experience (see Annex 1 for a complete list of participants).

The next step was to carefully select and invite a small group of graduate and postgraduate students to serve as teamwork discussion facilitators, as well as chapter development and follow-up leaders. This group was intended to be multidisciplinary in nature and ideally engaged in some capacity with the Blockchain@UBC cluster. An initial activity that the facilitators undertook was to review the papers that were prepared by the larger group of participants invited to the PWIAS RT Subsequently, a meeting was organized by the lead facilitators (i.e., Dr. Victoria Lemieux, the International Research Roundtable Co-Principal Investigator and Blockchain@UBC Co-Lead, and Dr. Marcelo Bravo, the lead International Research Roundtable Facilitator and expert in strategic design) to inform the teamwork facilitators of the preparations underway. The key information introduced and discussed during this preparatory time is included in the following figure (Fig. 1.3).

Part of the discussion at this meeting was to inform the future teamwork facilitators of this unique opportunity to bring world-renowned blockchain and DLT scholars to the University of British Columbia in order to co-generate knowledge together through discussion and peer-facilitated writing work. The following principles of the Strategic Design Method were shared and discussed:

1. Teamwork facilitators are expected to show initiative, and serve as mediators and catalyzers to move their theme group to achieve the specific day outcomes outlined by the leader facilitators;
2. Teamwork facilitators should pay attention to inviting all members of the team to participate, regardless of disciplinary background or area of expertise; and

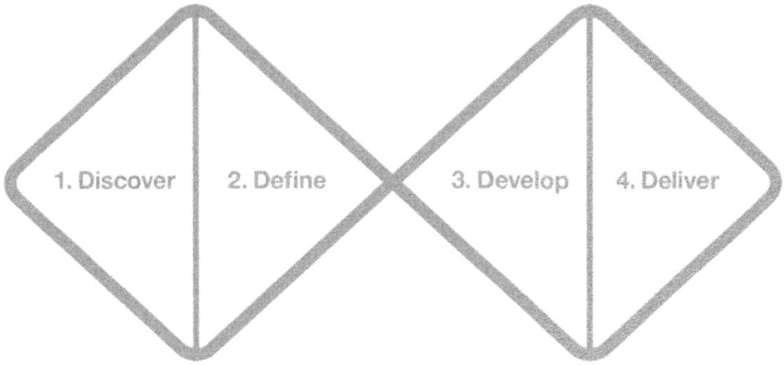

Fig. 1.4 The Double Diamond design process of the UK's Design Council

3. Teamwork facilitators are expected to guide and orient the group in a series of personal and group-shared reflections that will lead to written chapters in the following weeks/months.

A design model used to plan the collaborative experience with the teamwork facilitators relied upon an adaptation from the Double Diamond process developed by the Design Council in the UK. This process, according to the UK's Design Council (2005), has been used intensively in different realms of design challenges that could involve product design, service design, program development, etc. In this case, the lead facilitators used the model to portray a roadmap of the 2-day intensive co-reflection and co-writing work being planned. The four main elements of the Double Diamond were applied (see Fig. 1.4 for a high-level diagram of this model). It is important to note that the model is sequential but not linear in the sense that the four phases can sometimes overlap and inform the design process interactively and iteratively to achieve idea refinements and advancements.

First the *Discover* phase involved having the participants explore and self-select a theme from a broader list of potential themes. These themes reflected the intellectual contributions of participants, and served as early exploration entry points for the theme under discussion through the three-layer blockchain and DLT model reflection and their "*trade-offs in Socio/Economical and Political (Human interaction)—Data/Records (Recorded facts about human interactions that can serve as evidence), and Technical (Means of recording facts about human interactions*" (Lemieux 2019).

For the second phase known as the *Define* phase, participants were invited to focus and narrow down the types of reflections, ideas, and applied cases to be further explored in relation to the selected theme. The aim of this phase was to converge into a series of early peer-reviewed reflections.

The third phase, *Develop*, corresponded in this case to a process of collaborative, phased writing. In the first phase, participants were asked to individually write up their reflections on the selected theme. This was followed by team writing supported by the teamwork facilitator. This third phase was also programmed to include a

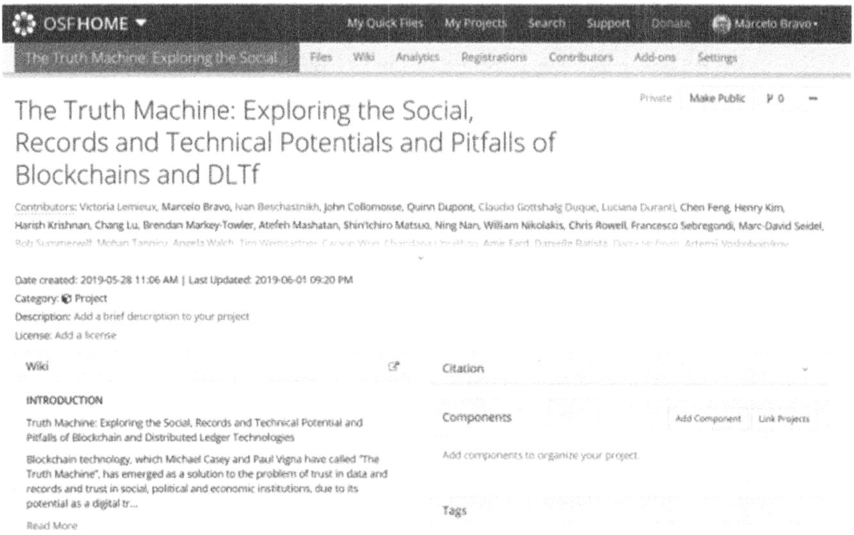

Fig. 1.5 The PWIAS RT's Truth Machine wiki page

mechanism of formal peer review, wherein experts from one chapter served as reviewers for other chapters.

Finally, the fourth phase *Deliver* was projected to be the final (off-site) write-up period, with suggestions and edits on draft chapters following a more traditional single blind peer-review process culminating in the final chapters to be included in the volume.

A final instruction provided to the teamwork facilitators was to familiarize themselves and plan to support participants with the Open Science Framework (OSF) dedicated space (see Fig. 1.5). The OSF is an open digital platform that provides a dedicated site for multi-user scientific communication that facilitates data sharing, storage of materials, feedback, and the creation of virtual collaboration groups. This platform was selected as a preliminary tool for participants' and facilitators' early exchange of ideas and reflections.

1.5 The PWIAS RT Experience and Follow-Up Collaborative Work

The PWIAS RT followed the Blockchain@UBC Annual Conference, which consisted of a full day of paper presentations by many of the PWIAS RT participants, along with introductions, discussions, and conversations with industry and community partners and the general public. This event, held at the University of British Columbia's downtown Vancouver campus at Robson Square, served as an opportunity for the participants to physically meet and discuss preliminary ideas on the themes to be considered for discussion.

As noted, the PWIAS RT comprised two full days of group work, initially involving the group at large, but then using a process of self-selection to form smaller working groups based on focal themes. Participants participated in the smaller groups based on their preference and expertise (see Annex 2 for a detailed agenda).

Day 1 of the PWIAS RT opened with the program agenda and broader introductions. Facilitators welcomed participants and informed them of the general plan for the two intensive reflection and writing days. Participants were given a document providing an overview of the roadmap of the experience, which included the double diamond process as well as the opportunity/challenge that participants were to be part of, stated as: *To co-generate knowledge and capture the interrelationships among the three layers in the design of blockchain and distributed ledger technologies: social, data/records, and technical ones—through a process of peer-led generative dialogue and group writing.*

An important component of the first morning of work was to inform the participants that the lead facilitators planned a design-led experience informed by the principles of strategic design. The facilitators explained that the strategic design perspective works at its best when multidisciplinary groups are invited, which was the case; where the problem, quest or challenge is broader than what can be solved through the lens of one academic discipline, which was also true; and importantly that strategic design was ideally practiced in studio settings. In order to achieve this last criterion, participants were introduced to a studio "etiquette" that required they:

- Internalize that the room (working space) was meant to replicate a studio experience, therefore active participation with both analytical and creative approaches were expected and encouraged.
- Contribute to the wider group discussion at large, as well as the dedicated small teamwork tasks, consequently exercising effective listening as well as disposition to participate in critical and creative thinking tasks.
- Refrain from early judgement that could halt the generative process of ideation and reframing.
- Be both knowledge "sharers" and knowledge "learners", and aim to enrich their unique disciplinary experience and practical expertise through the process of dialogue and active participation.

After this introduction, a follow-up activity conceived of as a "warm up" exercise was delivered by one of the facilitators. This group activity involved physical movement that required finding interesting facts about people's interests and the experiences of participants in the room. In order to do this, participants were required to find out information from their peers, and to obtain their signature as a means of verification for subsequent rewards (somewhat modeling the operation of blockchain and DLT systems). Typical questions included for example: find a participant that speaks at least three languages; find someone who has lived in Vancouver for at least 10 years; find someone who knows how the theme "Truth Machine" originated, etc.

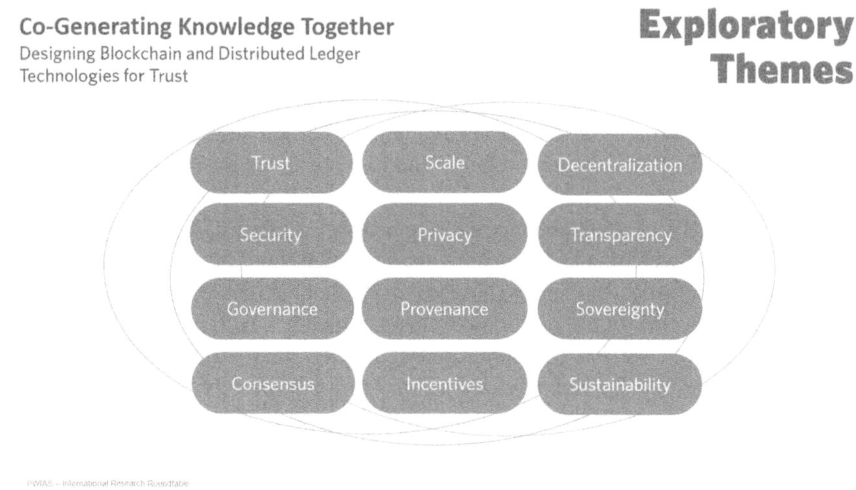

Fig. 1.6 The list of preliminary PWIAS RT themes for discussion

The idea was for participants to start engaging with each other in a friendly and collaborative way.

Following this active introductory team building exercise, participants were asked to focus on the challenge, i.e., how to co-generate knowledge together, based on the exploration and selection of pre-informed themes. To accomplish this, a list of pre-selected potential themes for further exploration was displayed by the facilitators (see Fig. 1.6 for the list of pre-selected themes).

This preliminary list constituted an integration of themes proposed originally by Dr. Victoria Lemieux, Blockchain@UBC's Cluster Co-Lead. This list of themes was the output of analysis of academic papers sent by PWIAS RT participants and early consultation with blockchain and DLT theoretical experts.

The next activity planned was the participant's selection of the theme of interest. For this, the facilitators converted the room into a physical open canvas where the name of each theme was included at the top of a large sheet of paper affixed to the walls. Participants were then asked to include early ideas and reflections on each theme. To activate this, participants were handed several "Post-it® notes" on which they could write down their ideas and affix them to the sheets of paper on the wall. The aim of this exercise was for the participants to contribute as much as possible to every theme/sheet, including adding to an unnamed sheet that was available for themes not previously considered (Fig. 1.7).

After this activity was performed, the facilitators requested that individual participants reflect and choose a theme of preference. Here, the idea was to have each participant mapped to each theme and select the most popular themes (i.e., those with the greatest number of ideas/reflections/Post It Notes). In addition to this self-selection process, the composition and size of the small working groups were taken into account. For example, the facilitators aimed to ensure that each group

Fig. 1.7 Participants contributing to "themes" development

included individuals whose expertise covered the three layers of the blockchain design space we were exploring. Ideally each small teamwork group was expected to include four to five participants of diverse expertise, as well as a graduate student serving as facilitator. The themes selected at the end of this process were: *Decentralization, Governance, Incentives, Provenance,* and *Security*, which have come to form the multi-authored chapters within this volume.

Subsequently, participants were introduced to the Open Science Framework (OSF) page or "Wiki" prepared for the occasion, as well as the writing protocol suggested for experts' written reflections and ideas. Facilitators then proposed the following structure for a group collaborative writing exercise: (1) to start with 20 minutes of individual writing, this writing to be recorded directly into the OSF wiki platform; then, (2) to stop writing and dedicate 20 minutes to read the comments and ideas expressed by their team members; and (3) to subsequently dedicate 20 minutes to offer a written response to colleagues' ideas, or to start a process of written exchange of ideas indicating similarities and differences in perspectives, as well as the need to conduct further analysis or research.

The first day concluded with two cycles of this collaborative writing process paused only by a planned small break as intermission. During this time, the facilitators were in direct contact with the participants in order to clarify expectations of the writing circles, as well as to hear direct feedback from the participants and the teamwork facilitators. This was an applied exercise of the practice of reflection, proposed by the Strategic Design Method that requires an openness to update or change processes for best results. At the end of the day, facilitators allowed for the small group facilitators to start customizing the writing experience for the purpose of

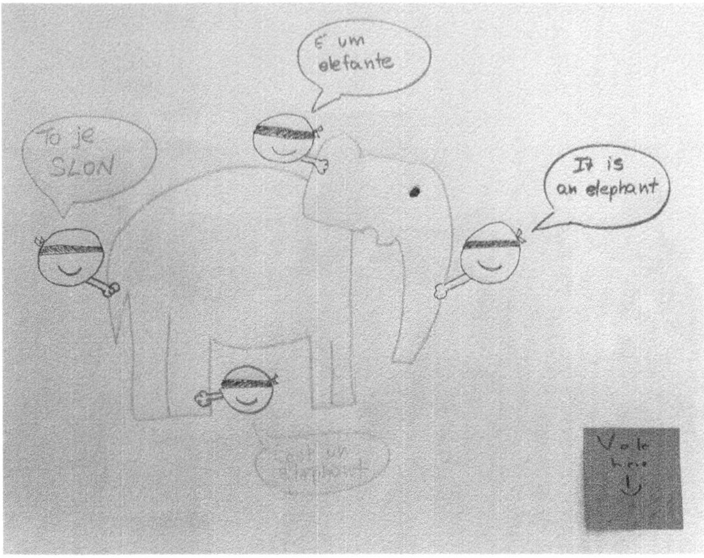

Fig. 1.8 The Truth Machine, as elaborated by PWIAS RT participants

increased team productivity and to allow a certain degree of team agency into the process as the teams became more comfortable working together.

Day 2 began with a warm-up that consisted of a creative visualization of the theme "Truth Machine". Participants were primed for this group task by inviting them to be as creative as possible and asked to come up with a designed scheme, picture, or representation of this abstract theme. Groups were also required to share their ideas with the larger group. During the day, and at break time, participants were requested to vote for the best drawing and members received a symbolic reward for this creative effort.

Figure 1.8 shows the winner of the "Truth Machine Illustration" exercise, elaborated by the Provenance team members.

The main purpose of this day 2 exercise was to continue advancing on the first day's ideas and reflections, and for the small groups to now work on a "Chapter visualization". To do this, groups were allotted a certain amount of time and were required to discuss, through reflection on their ideas, how their ideas could be refined during the overall writing time of the project. Participants were encouraged to use big charts where they visualized "how" the chapter might be integrated. It was required for them to be as visual as possible, to include the main elements of a possible chapter, and to always give consideration to the idea that the theme chosen had to discuss the interconnection of the three blockchain design layers that were at the core of the PWIAS RT discussion experience.

After the groups performed this activity, the lead facilitators requested that the groups share with the larger group the result of this activity. The objective of this task was twofold: to keep organizing the team's shared thinking on the chapter and to

Fig. 1.9 Small teamwork groups sharing their ideas of chapter visualization

open the floor for feedback and open reflection by other participants who, although not writing on that specific theme, were also knowledgeable on the subject and thus able to contribute with ideas and suggestions to aid in the preparation of chapters.

Figure 1.9 depicts the moments where participants were sharing and adopting feedback presented by the group at large.

Day 2 was thus proposed as an opportunity for small group integration and enhancement of the ideas and reflections on the chosen themes to be incorporated into this volume. It also included some opportunity for hearing about professional development opportunities in the blockchain and DLT domain: Participants informed one another about upcoming conferences, additional writing opportunities, research funding opportunities, as well as future plans for Blockchain@UBC as a multi-stakeholder cluster initiative.

Finally, the day ended with the review of the schedule for completion of the upcoming off-site work time, discussion of the type of support required from the teamwork facilitators, and discussion of the best ways for continuing communication and interaction. Participants were informed of the required commitment expected during the off-site time, which was articulated as the time for realization of the *Develop* and *Deliver* phases of the design-led process.

Notably, the time after the face to face PWIAS RT experience was identified as a critical time for the teamwork facilitators. This small group comprised of five graduate and postgraduate students served as a connector, guide, and in some cases, leaders for the successful completion of the remaining work. During this time, Dr. Victoria Lemieux had constant communication with these facilitators, supported the process of sending reminders, and was available to facilitate the follow-up work that resulted from the chapter drafts.

1.6 Applied Reflection, Lessons Learned from the Process, and Future Applications

The application of a design-led pedagogical process known as strategic design turned out to offer a useful and novel mechanism for participants' engagement in a designed sharing, reflecting and writing process. For instance, it confirmed the need and opportunity to incorporate both "face to face" academic meetings with the support of new "collaborative technologies" that facilitate project development. It was also clear that the particularities of and diverse composition of the cluster (e.g., the place of origin of participants) necessitated a design journey that included a fast and well-developed roadmap for participants' engagement and full understanding of the task at hand.

Another important lesson was the need to include constant check-ins with both the group at large as well as the teamwork facilitators. This constant feedback throughout both days, and beyond, allowed small yet fundamental adjustments to the pace and effective work of the small groups. Another interesting validation was the opportunity to test out the timing of the method, at least during the 2 days of the PWIAS RT. It was clear that the right combination of personal work time and group work time was an important element for participants to fully experience the process and to remain engaged.

Preliminary feedback also spoke to the balanced nature of including both critical thinking approaches as well as creative ones. As a premise of strategic design, the time dedicated to carefully plan and roll out the activities with the participants served to create an atmosphere of innovativeness in the process that aimed at materialized theoretical developments. It was also demonstrated that the inclusion of the above-described activities reinforced the nature of collaboration and opened space for creativity that was required as a means to meet the broader objective of the PWIAS RT: a multidisciplinary collaboration that involved a three-layered socio-informational-technical analysis.

Interestingly, this design-led model pointed to the importance of taking risks and the challenges of applying new methodologies that can support the goal of multidisciplinary collaborations. Although greatly appreciated in academia, multidisciplinary work is sometimes a common aspirational place or an ambitious objective that is difficult to realize in practice. Systemic barriers that prevent cooperation are still found, and key barriers exist at the epistemological and methodological level. The PWIAS RT confirmed that the Strategic Design Method can be used to overcome these barriers.

However, as with any other method, there is opportunity and invitation to continue with the development of new applications, tools, and techniques that can be applied in different contexts. The opportunity is there for a continued validation and experimentation of both strategic design as a method that supports and strengthens multidisciplinary collaborations, as well as the blockchain and DLT system theoretical advancements that emerged from the application of strategic design in the context of our International Research Roundtable.

Annex 1: Participants by Last Name

1. Beschastnikh, Ivan
2. Collomosse, John
3. DuPont, Quinn
4. Duranti, Luciana
5. Feng, Chen
6. Gottschalg Duque, Cláudio
7. Lemieux, Victoria
8. Kim, Henry
9. Krishnan, Harish
10. Lu, Chang
11. Markey-Towler, Brendan
12. Mashatan, Atefeh
13. Matsuo, Shin'ichiro
14. Nan, Ning
15. Rokmaniko, Maksym
16. Rowell, Chris
17. Sebregondi, Francesco
18. Seidel, Marc-David
19. Skwarek, Volker
20. Stancic, Hrvoje
21. Summerwill, Bob
22. Tanniru, Mohan
23. Tseng, Francis
24. Unnithan, Chandana
25. Walch, Angela
26. Weingärtner, Tim
27. Woo, Carson

Teamwork Facilitators

28. Batista, Danielle
29. Fard Bahreini, Amir
30. Hofman, Darra
31. Lu, Chang
32. Rowell, Chris
33. Voskobojnikov, Artemij

Pedagogical Facilitators

34. Bravo, Marcelo
35. Lemieux, Victoria

Annex 2: Agenda of Roundtable

June 11th, 2019

Time	Activity	Description	Location
9:00 am to 12:30 pm	Morning	9:00–9:30—Open and welcome 9:30–10:00—The three layers of blockchain design—Chris Rowell 10:00–10:30—Warm-up visualization exercise—Victoria Lemieux 10:30–11:00—Break 11:00–12:00—Design trade-offs idea generation	PWIAS Seminar Room
12:00 pm to 1:00 pm	Lunch	Sage Catering	
1:00 pm to 5:00 pm	Afternoon	1:00–3:00: Group writing session 3:00–3:30: Break 3:30–4:30: Group writing session 4:30–5:00: Preparation for Day 2	PWIAS Seminar Room

June 12th, 2019

Time	Activity	Description	Location
9:00 am to 12:00 pm	Morning	9:00–10:30—Group chapter visualization exercise 10:30–11:00—Break 11:00–12:30—Chapter structure planning	PWIAS Seminar Room
12:30 pm to 1:30 pm	Lunch	Sage Catering	
1:00 pm to 5:00 pm	Afternoon	1:00–3:00—Collaborative Group Writing and Editing 3:00–3:30—Break 3:30–4:30—Presentations on group progress and feedback 4:30–5:00—Workshop Close	PWIAS Seminar Room

References

Beausoleil, A. M. (2016). *The case for design-mediated innovation pedagogy* (Doctoral dissertation) Retrieved October 13, 2018, from http://circle.ubc.ca

Beausoleil, A. (2018). Why designers have arrived in corporate boardrooms. *The Conversation*. Retrieved November 15, 2018 from https://theconversation.com/why-designers-have-arrived-in-corporate-boardrooms-106437

Bellefontaine, T. (2013). *Innovation labs: Bridging think tanks and do tanks*. Ottawa: Policy Horizons Canada. Retrieved October 12, 2018 from http://www.horizons.gc.ca/eng/content/innovation-labs-bridging-think-tanks-and-do-tanks

Brand, W. (2017). *Visual thinking: Empowering people & organizations through visual collaboration*. Amsterdam: BIS.

Bravo Chapa, M. E. (2019). *Co-designing a university-wide framework: Structure, systems and services that support knowledge mobilization at UBC* (Doctoral dissertation). Retrieved August 31, 2019, from http://circle.ubc.ca

Brown, T. (2009). *Change by design: How design thinking transforms organizations and inspires innovation*. New York: Harper Business.

Brown, T., & Wyatt, J. (2010). *Design thinking for social innovation*. World Bank: Development Outreach.

Casey, M. J., & Vigna, P. (2018). *The truth machine: The blockchain and the future of everything*. New York: St. Martin's Press.

Cheney, J., Chiticariu, L., & Tan, W. C. (2009). Provenance in databases: Why, how, and where. *Foundations and Trends® in Databases, 1*(4), 379–474.

Collomosse, J., Bui, T., Brown, A., Sheridan, J., Green, A., Bell, M., Fawcett, J., Higgins, J., & Thereaux, O. (2018, August). ARCHANGEL: Trusted archives of digital public documents. In *Proceedings of the ACM Symposium on Document Engineering 2018*. New York: ACM

Design Council. (2005). *Double diamond design process model*. Retrieved August 30, 2019 from https://www.designcouncil.org.uk/sites/default/files/asset/document/ElevenLessons_Design_Council%20(2).pdf

Dunne, D. (2018). *Design thinking at work: How innovative organizations are embracing design*. Toronto: University of Toronto Press.

Edelman. (2017). *2017 Edelman Trust barometer: Global report*. Retrieved from https://www.edelman.com/research/2017-edelman-trust-barometer

Galsworth, G. D. (2018). *Visual workplace visual thinking: Creating enterprise excellence through the technologies of the visual workplace* (2nd ed.). Boca Raton, FL: CRC Press.

Julier, G., & Kimbell, L. (2016). *Co-producing social futures through design research*. University of Brighton. Retrieved from http://protopublics.org

Kannengiesser, N., Lins, S., Dehling, T., & Sunyaev, A. (2019). What does not fit can be made to fit! Trade-offs in distributed ledger technology designs. In *Proceedings of the 52nd Hawaii International Conference on System Sciences*. Available at SSRN: https://ssrn.com/abstract=3270859 or https://doi.org/10.2139/ssrn.3270859.

Kasireddy, P. (2018, February 3). EL15: What do we mean by "blockchains are trustless"? *Medium*. Retrieved from https://medium.com/@preethikasireddy/eli5-what-do-we-mean-by-blockchains-are-trustless-aa420635d5f6

Kiayias, A., & Panagiotakos, G. (2015). *IACR Cryptology ePrint Archive, 2015/1019*. Retrieved from https:/eprint.iacr.org/2015/

Kimbell, L. (2014). *The service innovation handbook: Action-oriented creative thinking toolkit for service organizations*. Amsterdam: BIS.

Kolko, J. (2012). *Wicked problems: Problems worth solving; A handbook & a call to action*. Austin: ac4d.

Lemieux, V. L. (2019). *The Truth Machine: Exploring the social, records and technical potential and pitfalls of blockchain and distributed ledger technologies – Introduction*. Retrieved from https://blockchain.pwias.ubc.ca/

Lemieux, V. L., Hofman, D., Batista, D., & Joo, A. (2019). *Blockchain technology & recordkeeping. ARMA International Education Foundation*. Retrieved from http://armaedfoundation.org/wp-content/uploads/2019/06/AIEF-Research-Paper-Blockchain-Technology-Recordkeeping.pdf

Liedtka, J., Ogilvie, T., & Brozenske, R. (2014). *The designing for growth field book: A step-by-step project guide*. New York: Columbia University Press.

Liedtka, J. (2017). Evaluating the impact of design thinking in action. In *Academy of Management Proceedings* (Vol. 2017, No. 1, p. 10264). Briarcliff Manor, NY 10510: Academy of Management. https://doi.org/10.5465/AMBPP.2017.177.

Martin, R. (2009). *The design of business: Why design thinking is the next competitive advantage*. Boston: Harvard Business Press.

Martin, R., & Christensen, K. (Eds.). (2013). *Rotman on design: The best on design thinking from Rotman magazine*. Toronto: University of Toronto Press.

Quayle, M. (2017). *Designed leadership*. New York: Columbia University Press.

Sniukas, M., Lee, P., & Morasky, M. (2016). *The art of opportunity: How to build growth and ventures through strategic innovation and visual thinking*. Hoboken: Wiley.

UBC d.studio. (2015). *Home*. Retrieved September 12, 2019 from http://dstudio.ubc.ca/

UBC Policy Studio. (2017). *Resilient cities policy challenge*. Retrieved September 12, 2019 from https://sppga.ubc.ca/research-impact/policy-studio/

Chapter 2
Blockchain Governance: De Facto (x)or Designed?

Darra Hofman, Quinn DuPont, Angela Walch, and Ivan Beschastnikh

2.1 Introduction: De Facto Governance in Blockchains

> [B]lockchain technology is being lauded as transformative for every human practice that uses recordkeeping (so, all of them). [. . .] if blockchain technology ends up enabling our most fundamental social infrastructures, then the governance processes for creating, maintaining, and altering the technology deserve careful scrutiny, as they will affect the resilience of the technology, as well as any infrastructure that comes to rely on it. (Walch 2019a, p. 59)

Examining the governance of blockchain technologies is critical but challenging. As Quinn DuPont (2019, p. 197) writes, "Governance is the buzzword in blockchains today [. . .] However, governance is notoriously difficult to define

---

(x)or, also known as "exclusive or" is a logical operation that outputs "true" only when inputs differ (one is true, the other is false); (x)or emphasizes mutual exclusiveness in the sense of "A or B, but not A and B." In the case at hand, there will be governance of the blockchain—if not designed, then de facto.

D. Hofman (✉)
School of Information, University of British Columbia, Vancouver, BC, Canada
e-mail: dhofman@mail.ubc.ca

Q. DuPont
School of Business, University College Dublin, Dublin, Ireland
e-mail: quinn.dupont@ucd.ie

A. Walch
School of Law, St. Mary's University, San Antonio, TX, USA

University College London, London, UK
e-mail: awalch@stmarytx.edu

I. Beschastnikh
Department of Computer Science, University of British Columbia, Vancouver, BC, Canada
e-mail: bestchai@cs.ubc.ca

V. L. Lemieux, C. Feng (eds.), *Building Decentralized Trust*,
https://doi.org/10.1007/978-3-030-54414-0_2

let alone operationalize. A definition for governance might be: stewardship, a mechanism that sets institutional rules and incentives, or the strategic exercise of power".

Governance, as traditionally understood, initially received little attention in the world of blockchain, at least outside of the technical dimensions of blockchain systems and their incentives. This occurred, in part, because of quintessential beliefs about blockchain technologies, at least in the public, permissionless form that has most captured the public imagination, such as Bitcoin and Ethereum. For example, an exclusive focus on the protocols bred a belief that blockchains are apolitical—"beyond the scope of governments, politics, and central banks" (De Filippi and Loveluck 2016, p. 1)—and that "algorithms are more trustworthy and authoritative than existing institutions," (Lustig and Nardi 2015, p. 747), a technocratic approach that "tries to solve issues of social coordination and economic exchange by relying, only and exclusively, on technological means" (De Filippi and Loveluck 2016, p. 1).

In other cases, such as "The Decentralized Autonomous Organization" (DAO), there was deliberate experimentation, an attempt to "create a social and political world quite unlike anything we have seen before" (DuPont 2018, p. 157). In most cases, however, early discussion about blockchain governance focused primarily on the technical aspects of the systems.

2.2 The Case for a Grounded Theory of Blockchain Governance

Given this context, there is relatively little literature on the prescriptive governance of blockchain platforms.[1] Consider, for example, the questions grounding Beck et al.'s blockchain governance framework: it becomes clear that very basic questions of governance ("How are decisions made?") remain open in the blockchain space (2018). However, while Beck et al.'s agenda is helpful, it is grounded in and informed by a theoretical framework of IT governance, which understands governance through decision rights, accountability, and incentives, and relies on agency theory.

Beck et al.'s work is certainly not the only lens through which to understand blockchain governance. De Filippi and Loveluck draw upon internet governance, by which they understand the internet as "a complex and heterogenous socio-technical construct [that] combines many different types of arrangement—involving social norms, legal rule and procedures, market practices and technological solutions—which, taken together, constitute its overall governance and power structures" (2016, p. 24). Walch (2019a) examines "decentralization" and the governance of blockchains through the lens of fiduciary law and the legal scholarship thereof,

---

[1]Some examples of work that discuss blockchain governance prescriptively include DuPont (2019) and Hofman et al. (2019).

while Hofman et al. (2019) discuss blockchain systems and the European Union's General Data Protection Regulation through the lens of information governance, informed largely by the lens of archival science. Not one of these approaches speaks to the totality of governance; each author takes a different approach to serve a different purpose.

Even though this literature attempts to situate blockchain governance within broader, mostly disciplinary, governance frameworks, the majority of existing literature on blockchain governance is descriptive and atheoretical, having largely arisen out of real-world crises of governance. The DAO hack of 2016 led to extensive analysis of governance challenges, in part because DAOs—decentralized autonomous organizations—are experiments in an entirely novel form of human governance (DuPont 2019; Walch 2019a). Seemingly prosaic or purely technical matters, however, have also led to crises of governance. De Filippi and Loveluck, in their examination of Bitcoin XT and the subsequent controversy over block size, note that "[t]o many outside observers, the contentious issue may seem surprisingly specific [. . ., but it] eventually led to a full-blown conflict which has been described as a 'civil war' within the Bitcoin community" (2016, p. 11). Ultimately, these crises have encouraged communities to find resolution not through code, but through social negotiation, or, in the case of "hard forks," the creation of new communities.

What has become clear from these crises is that while blockchains may permit experimentation with new forms of governance, they are not beyond or outside governance. After all, "[governance in] its purest form [. . .] describes the structures and decision-making processes that allow a state, organization or group of people to conduct affairs" (Bruce-Lockhart 2016). Even if it were possible[2] to set up a completely autonomous system of algorithmic authority in which all governance and management were executed on-chain, the structures and decision-making processes themselves would have to be agreed upon, created, and instantiated. Furthermore, this seemingly "autonomous" organization would still have to interact with the broader world. As De Filippi and Loveluck observe, "one cannot get rid of politics through technology alone, because the governance of a technology is—itself—inherently tied to a wide range of power dynamics" (2016, p. 16). For this reason, it may make more sense to adopt a grounded approach to development of governance theory for blockchains, rather than attempting to apply existing theories of governance to these novel contexts. Such a theory would take into consideration the social, institutional, and political contexts of blockchains, where these contexts are considered an essential part of understanding blockchain governance.

[2]It's not.

2.3 Situating Blockchain Governance in Existing Power Structures

> Decentralization inherently affects political structures by removing a control point [. . .] as Bitcoin evolves—and in the eventuality that it gets more broadly adopted—it will [. . .] encounter a variety of social and political challenges—as the technology will continue to impinge upon existing social and governmental institutions, ushering in an increasingly divergent mix of political positions. (De Filippi and Loveluck 2016, p. 15)

Blockchain protocols take their action within the existing world of material constraints, institutions, cultures and norms, and above all, existing sovereign governance systems. Decisions about the governance of any given blockchain system will impact and be impacted upon by these existing power structures: actions taken by participants within blockchain systems that violate nation state laws will be subject to state-based consequences. On the other hand, participants within blockchain systems continue to avail themselves of remedies offered by state actors (e.g., bankruptcy, fraud claims).

A substantial amount of rhetoric around blockchain technologies focuses on "decentralization" and "trustlessness." By enabling decentralized transactions and decision making and reducing or even eliminating the need to depend on humans, blockchains—or so the argument goes—will revolutionize how people interact, conduct business, and even govern themselves. Indeed, De Filippi and Loveluck describe the "implicit political project" of Bitcoin as "getting rid of politics by relying on technology" (2016, p. 22). In reality, however, blockchain technologies are complex sociotechnical systems, or as we argue in this volume, socio-informational-technical systems. "Decentralization" and "trustlessness" are both fraught terms that capture technical and social discourses and their interrelationships—a *promotion* of a kind of reality as much as a *description* of it.

As Walch explains:

> the term 'decentralized' is generally being used to describe how power operates in blockchain systems—suggesting that power exercised by people in these systems is diffuse rather than concentrated. This is critically important, as our understanding of how power is exercised within these systems will shape conclusions about how responsibility, accountability, and risk should work for them (2019b, p. 40)

Walch traces two major uses of "decentralization" in the discourse surrounding blockchain technologies, which are often conflated with one another: decentralization as a description of the network architecture which supports the blockchain, and decentralization as a description of "how power or agency works within permissionless blockchain systems" (2019b, p. 42). De Filippi and Loveluck similarly distinguish "between two distinct coordination mechanisms: governance *by* the infrastructure (achieved via the Bitcoin protocol) and governance *of* the infrastructure (managed by the community of developers and other stakeholders) (2016, p. 1). Even in Beck et al.'s study of Swarm City—a case study in which the interviewed developers have an explicit, ideologically-driven goal of making their code "increasingly decentralized and autonomous once it is implemented"—the developers nonetheless admit that "in order to make the tools, we initially need a really hierarchical governance," which they term a necessary "benevolent dictatorship" (2018, p. 1029). Technical decentralization can belie substantial centralization in how a system is actually designed and run, with tremendous decision-making power invested into the social structures surrounding the design, implementation, and operation of the system.

"Trustlessness" fares little better. Despite the fact that many of their participants used Bitcoin as "an act of resistance against institutions they felt had failed them" (Lustig and Nardi 2015, p. 762), Lustig and Nardi uncovered significant ways that participants relied on human judgement and trust. For example, they found that many of the individuals they interviewed spent 2–3 h per day trying to get informed about Bitcoin in order to learn "who to trust, how to protect their bitcoins from theft or fraud, and what community interventions were necessary to help Bitcoin itself run smoothly" (Lustig and Nardi 2015, p. 762). Similarly, DuPont found that, when The DAO's vision for novel governance broke down, people turned to "traditional models of sociality—using existing strong ties to negotiate and influence, argue and disagree" (2018, p. 2). Ultimately, De Filippi and Loveluck argue that, "although the *trustlessness* of the [Bitcoin] network seeks to obviate the need for a central control point, in practice, as soon as a technology is deployed, new issues emerge from unanticipated uses of technology—which ultimately require the setting up of social institutions in order to protect or regulate the technology" (2016, p. 25). Even "trustless" technologies, then, are connected to, protected and/or regulated by, and impact on social institutions of various degrees of trustworthiness.

"Decentralization" and "trustlessness," then, are not sufficient to exempt blockchains from governance, both internally (within the code) and externally (beyond the code). What that governance will look like, how blockchain governance will differ from other infrastructures, and how it will emerge, remains unknown. As Beck et al. note, "how exactly governance will change in the emerging blockchain economy is still little understood. Nevertheless, the promise of the blockchain economy is dependent on the implementation of effective governance mechanisms, which are, in turn, dependent on a thorough understanding of the phenomenon" (2018, p. 1029). Their study on IT governance, identifies a number of open questions for governance in what they term the "blockchain economy" (see Fig. 2.1).

2.4 Blockchain Governance Analysis Framework

> [G]overnance is [...] strategic and visionary. Governance involves the assessment of multiple options, limitations, and opportunities (DuPont 2019, p. 23)

Given the great variety of blockchain technologies, the myriad purposes to which those blockchains might be put, and the limitations of existing theoretical perspectives on blockchain governance, we take a step back and pose the following question as a guide: what ought to be a *theory* of blockchain governance, specifically, one that

Dimension	Research questions
Decision rights	• How are decisions made in the blockchain economy? • How are decision management rights and decision control rights allocated? • How is disagreement about decision-making resolved in the blockchain economy? • What is the role of ownership in the blockchain economy?
Accountability	• How is accountability determined in the blockchain economy? • How is identity engrained in the blockchain economy? • How is transaction enforcement embedded in the blockchain economy? • How are disputed transactions resolved in the blockchain economy? • How is trust affected by the blockchain economy? • What is the role of institutions in the blockchain economy?
Incentives	• How is consensus incentivized in the blockchain economy? • How does incentive alignment work in the blockchain economy? • How is system use incentivized in the blockchain economy? • How is system development and maintenance incentivized in the blockchain economy? • How do business models shape the blockchain economy?

Fig. 2.1 Research agenda for governance in the blockchain economy (Beck et al. 2018, p. 1029)

is endogenous to the socio-political, economic, cultural, informational and technical realities that define crypto? This is not a question of "what is or ought to be governance" but rather, what would or should a meaningful theory of crypto governance be, where "meaningful" means a theory that is analytically descriptive and prescriptive. Our framework is meant to enable descriptive or prescriptive analyses of blockchain platforms, acknowledging that, "there is no one right approach to [blockchain] governance [...] there are risks and opportunities for each" (DuPont 2019, p. 198).

Our framework, shown in Fig. 2.2, tries to capture the embeddedness of blockchain solutions in the broader world, noting that this is based on our review and understanding of the existing blockchain literature rather than a much needed rigorous grounded-theoretic analysis of blockchain governance.

We took the water cycle as an exemplar, where blockchain governance is a small part of much broader, more complex systems. Similar to the water cycle, blockchain governance exists within, is determined by, and ultimately determines the broader world in which it is embedded. The reciprocity in the framework—the "world" in our water cycle—captures the fact that blockchain systems do not exist separately from the broader world. "Even in a world with widespread use of blockchains, governments still retain their four regulatory levers—*laws, code, market forces, and social norms*—which could be used to either directly or indirectly regulate this new technology" (De Filippi and Wright 2018, p. 208). We add a much-needed fifth category—the environment—because environmental factors have a direct impact on our social and institutional systems broadly, and on all blockchain systems specifically.

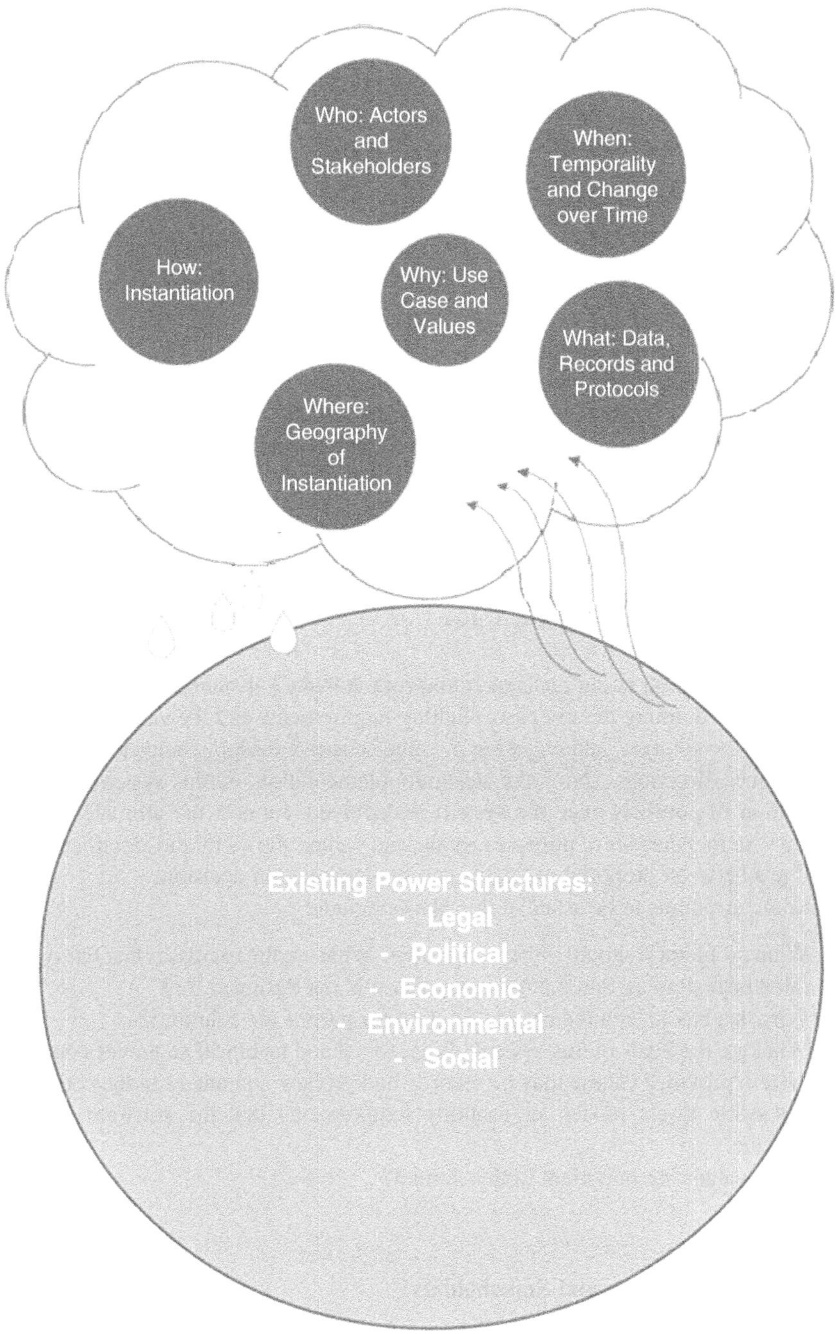

Fig. 2.2 Governance analysis framework

This framework is meant to serve as a high-level analytic tool; given the enormous variability in blockchain systems (including incommensurabilities such as values and norms), we propose an inclusive, question-led approach, which enables the examination of governance for any system without a priori prescribing technical goals or socio-economic realities.

2.4.1 *Within the Cloud: Internal Governance*

The "cloud" in our model represents governance modalities of the blockchain system itself. Governance of the blockchain system interacts with existing power structures in complex ways that co-determine each system's modalities. We imagine a homeostatic relationship between internal and external governance mechanisms. At the centre of the cloud—the origin of governance theory—is the question "why?" Governance choices flow from these exogenous values.

2.4.1.1 Why: Values and Use Cases

The initial question in our analysis framework is "why?" Establishing the "why" of the system—defining the use case, eliciting requirements and the values behind the design of the solution, and engaging in value-sensitive design—helps to ensure that governance decisions about the design/implementation of the system, and the resolution of conflicts once the system is deployed, support the ultimate purpose of the system. Analysis of purposes, goals, and values allows for the identification of conflicts between proposed use cases and implementation decisions.

Some questions to be asked at this phase include:

- What problem(s) should this system solve? What are the use-cases that the system intends to support, and the use-cases that it is not designed for?
- Why is a blockchain the chosen solution (or part of the solution)?
- What are the goals of this system? What social and technical guarantees does the system provide? (These may be security and privacy guarantees in the context of a specific threat model, or usability requirements that the software aims to provide.)
- What values are important in this system?

2.4.1.2 Who: Actors and Stakeholders

The next step is to identify actors and stakeholders and to identify their interests, rights, and obligations.

Some questions to ask at this stage include:

- Who are the actors and stakeholders (or better yet, "actants")? Identifying actors is often a practical challenge, especially when systems are designed to be privacy-preserving or purposefully obfuscate the actors. Some of the direct actors in a public blockchain system include developers, record producers, nodes, and the designers or creators of the system. As these systems integrate further into socio-economic infrastructures, this list and its complexity grows to include end users, public policy makers, and the broader ecosystem engaging with or building on the blockchain system.
- How are the actors in the system identified and how are their identities regulated? Public and private cryptographic keys, email addresses, names, and many other approaches may be used to identify and regulate participants. These design choices will constrain if and how actors may prove their identity, change or create new identities, leave the system, maintain anonymity, and so on.
- What expectations do we have of them? What actions will or might they take? How will these actions impact others?
- Will some actors act on behalf of others? On what (moral, legal?) ground do they implement the will of others? (Which others?)
- How is discretion exercised when conflict arises? When is consent, permission, and authority needed, granted, or assumed?
- What norms or other frameworks constrain the behavior of actors?
- What types of actions are forbidden, encouraged, or tolerated?
- What norms or other frameworks constrain the designers or creators of the systems?

Research and development norms and values deserve special mention here. In his study of research and development norms in the field, DuPont (2020) found that software developers are largely aware of formal guidelines but made little use of such guidelines: Perhaps most worrisome, DuPont found that researchers and developers have significant unacknowledged conflicts of interest, use risky research methods, and lack safe mechanisms for disclosure reporting. DuPont (2020) concluded that because these systems typically involve valuable tokens (for game-theoretical security models and decentralized funding structures), they comprise a new kind of per se value technology, with research and development governance challenges that rival bio- and nanotechnology.

Norms determine governance behaviours. For example, with developers, there may be norms determining that a developer will not try to thwart the system or that contributing to an open source software project is a virtuous act of contributing to the common good. Similarly, there may be norms around reputation—if a developer is seen to be trying to harm the system or seen to be incompetent, such behaviours will damage their reputation and future earnings. As such, these potential consequences may constrain governance options. The public nature of the software code also constrains a developer's behavior to some extent. Since code is subject to public scrutiny, bad/incompetent actions by developers will be revealed (assuming the veracity of "Linus's Law" that "given enough eyeballs, all bugs are shallow" (Raymond 1999)). Transparency of the code here is an "architectural" constraint on developer behaviour. However, people may not be able to read code, and typically in practice, relatively few people actually review code even when it is open source. Also, sophisticated developers may be able to hide actions in platform or contract code (even when surreptitious code injection is for the acknowledged benefit of the system, as has happened in the past, this capability introduces governance questions). The social aspects of blockchain governance, then, can be as complex and nuanced as the technical aspects.

2.4.1.3 When: Temporality and Change Over Time

As noted *supra*, blockchain solutions change—both in function and through their relationship to broader structures of power. This iterative relationship is why we chose a homeostatic model of governance for this framework. Thus, in determining the governance of the blockchain, questions of temporality and change over time—the system's lifecycle—must be asked, such as:

- How will governance address actors' changing relationships to the system over time? Are all developers fungible? Must the system be able to differentiate between different classes of records producers and users, and in what ways?
- What known future changes will the system have to be able to respond to? For example, if there are legal or regulatory changes, how will the system and its actors—including "autonomous" components—respond? Likewise, how will other risk factors be addressed, including those that lie unknown in the future and that may present existential or systematic risk?
- Could future events bring about consequences where the platform ought to be destroyed? Lifecycle management affects all system components, including assets no longer under control.

2.4.1.4 What: Data, Records, and Protocols

Blockchains serve to store and/or help protect the integrity of data and/or records. In order to understand and/or establish the governance of a particular blockchain solution, it is necessary to understand what that system stores and how it provides the intended functionality. The technical realization of a blockchain will simultaneously impose demands and constraints on the governance structure. For example, if the data is arbitrary and is stored without revealing the origin of the data, then governance must concern itself with issues like copyright infringement and whether or not to establish structures that would impose constraints on the data allowed into the system.

Questions to ask at this stage include:

- What data and/or records must the system store? (What are the legal or regulatory obligations?)

- What data and/or records must not be stored in the system? (For purposes of privacy, financial risk management, or corporate policy.)
- Are there data and/or records that require special consideration? For example, are there data and/or records containing personally identifiable information that requires special treatment under law?
- Are there data and/or records that must not be kept indefinitely?

2.4.1.5 Where: Geography of Instantiation

While blockchain solutions are largely treated as borderless in the popular imagination, state actors continue to exercise territorial (and extraterritorial) jurisdiction, even in cyberspace. As just two examples, the great firewall of China determines what internet content is accessible to people who access the internet from Chinese territory (Griffiths 2019), and ISPs in the USA distinguish between internet traffic between end-points that are both in the USA versus traffic where one of the end-points is outside of the USA (Gallagher and Moltke 2018; Goldberg 2017).

Furthermore, depending on the use case, being able to demonstrate compliance with laws and regulations may be necessary. And, beyond law and regulation, there are economic, political, social, and environmental constraints that are specific to their geography; e.g., a Proof of Work consensus mechanism might be prohibitively expensive in an area with high electricity costs (or, alternatively, in a very hot area where significant cooling would be required).

Some questions to ask about where a solution is instantiated:

- Are there any reasons why this solution must be instantiated in a particular location? For example, data localization laws might require data to be held in a particular legal jurisdiction (which limits both the "where" and the "how" of the instantiation).
- Are there any reasons why this solution should *not* be instantiated in a particular location?
- Are there location-based strengths/weaknesses that encourage adoption of a private blockchain instead of the broadly-distributed public blockchains?
- Is there a differentiation in access or power granted to actors in the system based on their geographical locale? For example, diversity of location (of nodes, users, etc.) may be encouraged and even required in systems that aim to avoid becoming too geographically centralized.

2.4.1.6 How: Instantiation

Finally, after establishing all of the above, governance must address executable code (the technical layer). Data and records are instantiated, but so are implicit, social properties that affect communities of developers, records producers, and users.

Some questions to ask about the instantiation of the solution include:

- What kind of blockchain solution best meets the governance needs of the system? Public, private, permissioned, permissionless?
- What technical features increase governance capacity?
- What consensus mechanism best meets the needs of both the use case and the actors?
- How will buy-in of the necessary communities be made clear?

2.5 Conclusion

> When it comes to freedom and autonomy, the assumption that the rule of code is superior to the rule of law is a delicate one—and one that has yet to be tested. (De Filippi and Wright 2018, p. 207)

Given the extensive role that blockchain technologies—and new blockchain-enabled forms of organization and interactions, such as DAOs—could play in society, we must consider governance of, by, and through blockchains to ensure that we identify areas of risk and in turn understand how conflict and crisis can be handled. By adopting a meta-theoretical model of homeostatic interaction, anchored in the values of a given set of actors, our framework proposes opportunities for innovation in governance. With their incentive and prohibition mechanisms, decentralized architectures, and ontologies of per se value, blockchain systems provide opportunities for social experimentation (DuPont 2019). "Governance" might indeed be difficult to define and operationalize, but trying to do so, through a grounded and contextual approach, is a necessary step to ensure that blockchain solutions can meet their potential.

References

Beck, R., Müller-Bloch, C., & King, J. L. (2018). Governance in the blockchain economy: A framework and research agenda. *Journal of the Association for Information Systems, 19*(10), 1. https://aisel.aisnet.org/jais/vol19/iss10/1

Bruce-Lockhart, A. (2016). What do we mean by 'governance'? *World Economic Forum.* https://www.weforum.org/agenda/2016/02/what-is-governance-and-why-does-it-matter/

De Filippi, P., & Loveluck, B. (2016). The invisible politics of Bitcoin: Governance crisis of a decentralized infrastructure. *Internet Policy Review, 5*(3). https://doi.org/10.14763/2016.3.427.

De Filippi, P., & Wright, A. (2018). *Blockchain and the law: The rule of code.* Cambridge, MA: Harvard University Press.

DuPont, Q. (2018). Experiments in algorithmic governance: A history and ethnography of "The DAO," a failed decentralized autonomous organization. In M. Campbell-Verduyn (Ed.), *Bitcoin and beyond* (pp. 157–177). New York: Routledge.

DuPont, Q. (2019). *Cryptocurrencies and blockchains.* Cambridge: Polity Press.

DuPont, Q. (2020). Guiding principles for ethical cryptocurrency, blockchain, and DLT research. *Cryptoeconomic Systems Journal.*

Gallagher, R., & Moltke, H. (2018). The wiretap rooms: The NSA's hidden spy hubs in eight U.S. cities. *The Intercept.* https://theintercept.com/2018/06/25/att-internet-nsa-spy-hubs/
Goldberg, S. (2017). Surveillance without borders: The "traffic shaping" loophole and why it matters. *The Century Foundation.* https://tcf.org/content/report/surveillance-without-borders-the-traffic-shaping-loophole-and-why-it-matters/
Griffiths, J. (2019). *The great firewall of China: How to build and control an alternative version of the internet.* London: Zed Books.
Hofman, D., Lemieux, V. L., Joo, A., & Batista, D. (2019). The margin between the edge of the world and infinite possibility. *Records Management Journal, 29*(1/2), 240–257. https://doi.org/10.1108/RMJ-12-2018-0045.
Lustig, C., & Nardi, B. (2015). Algorithmic authority: The case of Bitcoin. In *2015 48th Hawaii International Conference on System Sciences (HICSS)*, HI (pp. 743–752). Los Alamitos, CA: IEEE. https://doi.org/10.1109/HICSS.2015.95
Raymond, E. S. (1999). *The cathedral and the bazaar: Musings on Linux and open source by an accidental revolutionary* (Vol. 12, pp. 23–49). Cambridge, MA: O'Reilly.
Walch, A. (2019a). In code(rs) we trust: Software developers as fiduciaries in public blockchains. In I. Lianos, P. Hacker, G. Dimitripolous, & S. Eich (Eds.), *Regulating blockchain: Techno-social and legal challenges* (pp. 58–81). Oxford: Oxford University Press.
Walch, A. (2019b). Deconstructing 'decentralization': Exploring the core claim of crypto systems. In C. Brummer (Ed.), *Cryptoassets: Legal, regulatory, and monetary perspectives* (pp. 39–68). New York: Oxford University Press.

Chapter 3
Incentives to Engage Blockchain and Ecosystem Actors

Mohan Tanniru, Jianyu Niu, Chen Feng, Claudio Gottschalg Duque, Chang Lu, and Harish Krishnan

3.1 Introduction

The American Psychological Association (APA) defines "incentive" as an external stimulus, such as a condition or an object, that enhances or serves as a motive to influence behaviour (Incentive 2015). An incentive system is defined by a set of rules and rewards, dictated by the environment within which the value of the rewards to influence behaviour is perceived. In this chapter, we explore two aspects of the concept of incentives that are of particular interest at the present time in relation to blockchain. Firstly, we delve into the adoption of emerging technologies, like blockchain, considering incentives as focused on the value or relative advantage that technologies provide to improve the way we work and interact with each other, and what incentives enterprises may have to adopt them. Secondly, we consider incentives in the context of blockchain consensus mechanism design; that is, the

M. Tanniru
College of Public Health, University of Arizona, Tucson, AZ, USA
e-mail: tanniru@oakland.edu

J. Niu · C. Feng
School of Engineering, University of British Columbia (Okanagan Campus), Kelowna, BC, Canada
e-mail: jianyu.niu@ubc.ca; chen.feng@ubc.ca

C. G. Duque
Faculty of Information Science, University of Brasília, Brasília, DF, Brazil
e-mail: klauss@unb.br

C. Lu (✉)
Blockchain@UBC, University of British Columbia, Vancouver, BC, Canada
e-mail: chang.lu@ubc.ca

H. Krishnan
Sauder School of Business, University of British Columbia, Vancouver, BC, Canada
e-mail: harish.krishnan@sauder.ubc.ca

V. L. Lemieux, C. Feng (eds.), *Building Decentralized Trust*,
https://doi.org/10.1007/978-3-030-54414-0_3

rules and procedures by which nodes in a blockchain agree to record a transaction and update the distributed ledger.

These two aspects of incentives, which involve both developer/administrator and enterprise user communities, are often actually two distinct ecosystems, each with its own stakeholders. For example, *stakeholders within* a development/administrator ecosystem may be defined as those who create data/record blocks or set up consensus protocols to validate those blocks and allow external stakeholders (users of a blockchain ecosystem) to share resources. Similarly, *stakeholders of the enterprise ecosystem* may be those who communicate and coordinate their activities by sharing resources managed by external stakeholders (developers of a blockchain ecosystem). In some cases, these roles may be performed by the same individuals, but generally they are performed by different individuals.

We argue that if blockchain technology assimilation is to be effective, both of the above-noted aspects of incentives, and their respective ecosystems, must be considered in the design of blockchain ecosystems. We begin our chapter by laying out our understanding of the characteristics of a blockchain ecosystem.

3.2 The Blockchain Ecosystem

We depict a blockchain ecosystem on the left hand side of Fig. 3.1. The ecosystem has multiple stakeholders with varying roles: those involved in developing the technology platform or layer used to share resources; those who develop administrative protocols to ensure the integrity of those sharing and using resources through authentication and consensus protocols, and those who actually write the resources on to the network, often called miners, in immutable data/record blocks to support secure access.

The users/actors of the social/application layer, or dimension, are stakeholders of the enterprise ecosystem shown on the right hand side of Fig. 3.1. These are customers of the blockchain platform and may use their own application interfaces

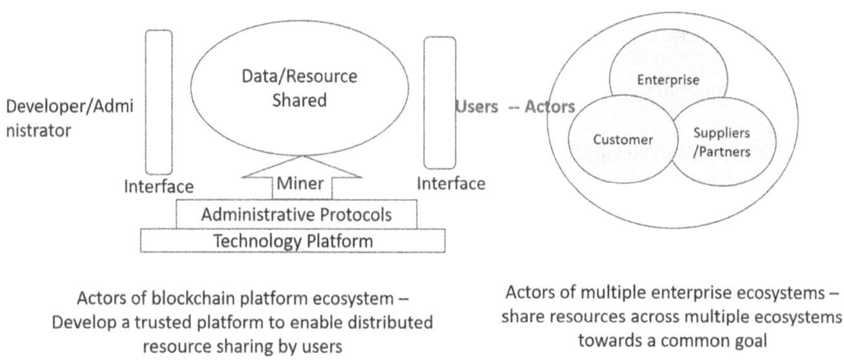

Fig. 3.1 A blockchain schema with multiple stakeholders

(apps or other systems within their ecosystems). Their ecosystem is often influenced by the goals of the enterprise that create value to these customers by leveraging their own resources as well as the resources of its suppliers/partners.

Information systems development research has used three different layers (architecture, data, and application) to recognize the need for independent development of each due to evolving technology and usage characteristics, while recognizing their interdependencies for ensuring implementation success. By separating the technology development from its use, the platform becomes "application agnostic" and supports generalizability. This allows each ecosystem to develop rules and norms that help govern their stakeholders and the resources of each ecosystem. Such a division of responsibility has been used to increase the agility with which each can make decisions to reflect changes within its own environment.

The technology layer, or dimension, in support of an enterprise ecosystem may have to reflect the characteristics of the application. For example, healthcare applications will continue to evolve with changes in regulations and the role of various stakeholders in the associated ecosystem in support of patient care (clinical and non-clinical care providers inside and outside hospital walls). This will potentially affect the way patient resources are created and used and by whom.

The technology and data/record layers/dimensions may have to be sensitive to the environments within which the enterprise operates. For example, the enterprise may be operating in a less technologically mature environment for supporting resource sharing (e.g., care delivery models in rural settings) or in countries where data standards for privacy and control are different and evolving (as in General Data Protection Regulation (GDPR) in Europe).

For these reasons, the blockchain architecture is represented along three dimensions or layers for independent development, while recognizing their dependency. The technology architecture has its own ecosystem with a platform, protocols and miners supporting resource sharing.

The data/record layer provides an opportunity to allow resources to be written and shared by the users of the blockchain for meeting the enterprise goals. The developers/administrators decide on the interface to allow miners to input records into the blockchain, and the users/actors of the enterprise ecosystem will use an interface to input and access resources needed for their application.

The social/application layer is implicit in the sense that enterprise actors/users are driven by their incentive to share resources to meet their application goals, and miners are driven by their incentives to reap benefits by adding the resources to the blockchain.

- **Technology architecture**—the architecture used (Ethereum, Hyperledger, etc.) to support resource sharing will continue to evolve
- **The data/record layer**—the standards for inputting, securing, and sharing records will continue to evolve
- **Social/application layer**—the incentives/business models will continue to evolve as the platform is used in multiple enterprise domains and across applications

Blockchain architecture platforms (permissionless or permissioned) are often developed and administered by one set of parties (left hand side of Fig. 3.1) and used by another set of parties, e.g., enterprises that pay for the services these platforms provide (right hand side of Fig. 3.1). In some cases, the people, or miners, who create data/record blocks that support resource sharing are either part of the creator/administrator (in permissioned blockchains) or are independent and need to be included in the business model (in permissionless blockchains). In a permissionless platform, miners must be incentivized.

3.2.1 Enterprise Ecosystems and Incentives for Assimilation of Blockchain Technology

3.2.1.1 Incentives, Motivation, and Information

Motivated people are proactive, persistent, and enthusiastic, and they normally know what they want and what is relevant to pursue (see Fig. 3.2). However, not all people are motivated, especially when the change called for seeks to alter institutional norms and practices with which they are comfortable. This reason for lack of motivation is the case with the assimilation of blockchain technology, where enterprise stakeholders are asked to trust the information they share to an architecture with no central coordination.

Relevance is key in motivating people to change and can act as an incentive. Relevance is needed both when information is presented to make people *aware of the technology and its capabilities*, and when information is used to *motivate people to change their behaviour*, if this is required. In the case of blockchain technology, relevance is needed to support two different stakeholder groups: developers, and enterprise users. Each has their own motivations and institutional context to work

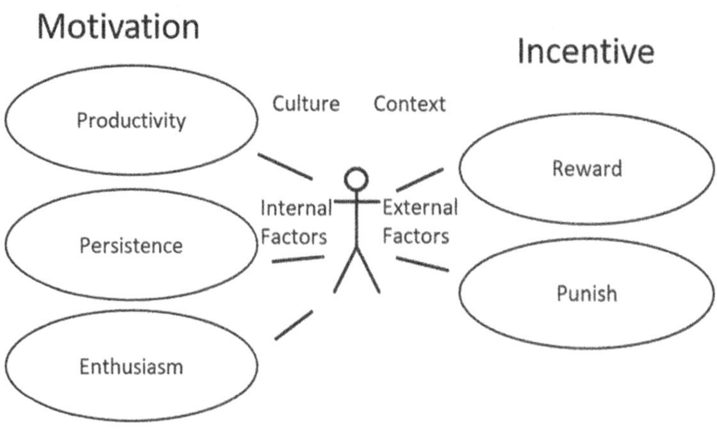

Fig. 3.2 Motivation and incentive

within, but they need to work towards shared goals if the technology is to reach its full potential.

3.2.1.2 Information Relevance to Support Awareness and Assess Capabilities

In the digital age, in which the exploitation of advanced technologies is paramount to creating value by adding services to address customer demands, developers and users search for information from different sources to become aware of what is possible, and they use an entrepreneurial mindset to synthesize this information to assess their capabilities. Engaging in conversations with others on social media, digital communities, and forums can help both developers and users become aware of the potential that advanced technologies have for value creation. Researchers at higher educational institutions often play a role in supporting awareness and enhancing capabilities by sharing research, supporting interaction among actors of both ecosystems, and using pilot applications to unearth both technology and individual capabilities.

As individuals search for information and make inferences about the trustworthiness of a distributed platform such as blockchain, they may need information from multiple communication channels. Information professionals, who are responsible for recording and disseminating advances in blockchain technology from academic and practitioner case studies, can play a major role in communicating effectively to those who want to understand concepts somewhat out of step with the norms used in today's practice, such as "miners" competing to add data and "trusting" platforms to share resources with no central coordination.

3.2.1.3 Information Relevance to Motivate Change Behaviour

Multiple theories have discussed the need for change (content theories), the value of change (process theories) and the motivational drivers of change (contemporary change). Addressing the *need for change, the value of engaging in change processes, and the driving forces that motivate one to change* calls for tailoring the information to influence the different stakeholders involved (Saif et al. 2012).

As technologies help to overcome barriers of space and time, collaboration among several actors from multiple ecosystems to share resources has become both viable and a competitive necessity. Justifying the *need for change* to act quickly in today's fast changing technology and enterprise landscape is not difficult, but it should be communicated effectively. Also, the pace needed to learn and apply new technologies calls for gaining insight from a vast amount of available information in a short time. The velocity of interactions (number of searches and interpretations sought through inferencing over a given time period) can be high given the rapid pace of change in the competitive landscape. Under these environments, internalizing one's understanding of the need for change and assessing its impact on one's own work processes is much harder. Information on the need for and impact of changes may need to be presented in different forms, backgrounds, shapes, etc., both to illustrate the benefits and contrast the narratives of pre- and post-change situations (as in blockchain implementations). To illustrate, Fig. 3.3 shows some architectural features discussed in information literature (Brandao and Duque 2011).

Lastly, drivers that motivate individuals to change their behaviour have to be contextualized to the role individuals play within their ecosystems (e.g., developer or user). For example, developers learn new technologies to improve their knowledge/skill and marketability, entrepreneurs seek new opportunities to develop commercially viable services, and enterprise users adopt technologies to reduce costs or build customer relationships. Incentives used to influence behavioural change need to take these individual drivers into consideration in order to drive the desired changes in behaviour.

In summary, human beings need incentives (external factors) to change their behaviour, and motivators (internal factors) to sustain behavioural change. Providing contextually relevant information with increased velocity (multiple interactions with keen insights in shorter time intervals) is needed to bring about change among stakeholders of both blockchain and enterprise ecosystems. The next few sub-sections will elaborate on incentives needed and strategies used to support change in each of these two ecosystems.

3.2.1.4 Enterprise Ecosystems and Platform for Distributed Data Sharing

Enterprises in the digital age are influenced by a complex environmental dynamic—empowered customers and emboldened technology entrepreneurs. The customers are empowered to demand services when and where they want them using personalized devices (e.g., smart phones, wearables) and communication tools (e.g., social media, information exchanges, wireless technologies). Technology entrepreneurs are emboldened with their ability to cater to customer demands quickly by leveraging their knowledge capital and evolving technologies. Both practitioner literature (Aghena et al. 2015; Bossert et al. 2014) and complexity theory research (Uhl-Bien et al. 2007) argue that enterprises need agility to operate at dual speeds to compete in the digital age. While the faster speed is needed to explore and evaluate innovative customer services using technology partners, the regular speed helps enterprises learn from exploration and adapt services shown to be viable into regular business to compete and grow.

Complex adaptive systems (e.g., biological, sociological, environmental) have used an instinct for self-preservation by learning to adapt to changes within and around their ecosystem and build resiliency. Enterprises, even if their natural instinct is for competitive survival, need to broaden their ecosystem to include customers and technology partners in today's digital age to build such resiliency. By engaging with actors and their resources from customer and partner ecosystems, enterprises can begin to build resiliency by expanding their capacity to absorb new ideas, adapt

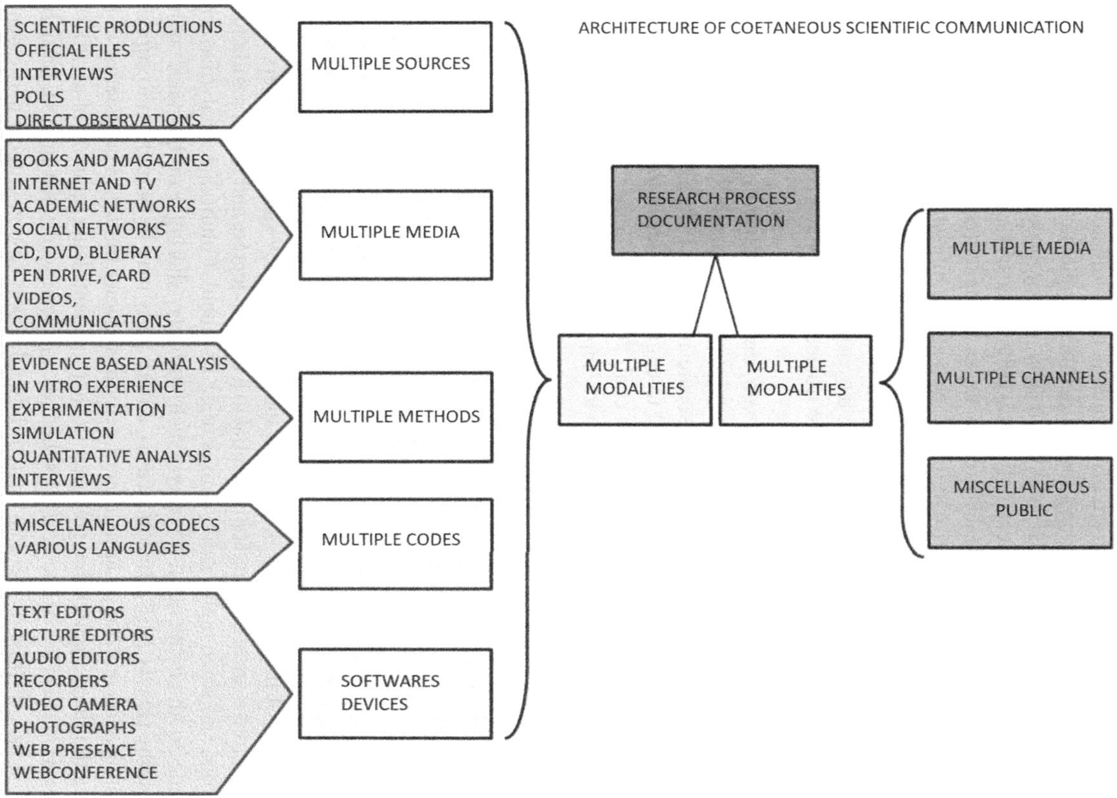

Fig. 3.3 Architecture of coetaneous scientific communication (Reproduced (in translation) from Brandao and Duque 2011, Fig. 1)

those shown to be viable, and transform the regular business by accommodating these ideas. In fact, complexity theory argues that a mix of administrative, enabling, and adaptive leadership styles are needed to build such organizational capacity (Marion 2008; Uhl-Bien et al. 2007). The goal is to let enabling leaders explore and evaluate new ideas, adapt leadership to learn and absorb these ideas, and administer leadership to transform organizations to assimilate viable ideas.

Research in service science and service dominant logic has emphasized the need for enterprises to use a broader ecosystem lens as they become service centric in value creation in today's increasingly knowledge-intensive market (Spohrer et al. 2008; Vargo and Lusch 2006). Use of the actors and resources of multiple ecosystems will help improve enterprise agility (Lusch and Nambisan 2015), but this poses significant challenges, since each ecosystem has its own norms and practices that guide the way actors integrate resources to meet their goals (Wieland et al. 2018). Hence, any platform designed to support the communication and coordination of actors across multiple ecosystems must be trusted by all involved, even if they all share the same goals.

Since the introduction of the Internet/web, enterprises have begun to use platforms to communicate with peers and customers and coordinate activities with other enterprises (suppliers and partners) with secure intranets. Platforms such as social media and digital infomediaries were used to engage customers to gain insight into their needs and answer questions (Khuntia et al. 2017). Still other platforms such as digital exchanges were used to interact with suppliers to share information, even though these have yielded limited success, especially when they are viewed as untrustworthy in supporting shared goals (Gerst and Bunduchi 2006). Some enterprises built platforms as their business model to support peer-to-peer communication and resource sharing (Eisenmann et al. 2009), while others have broadened their e-commerce platforms to the activities of others, such as logistics (e.g., Amazon), or the creativity of others in software product development (e.g., Apple, Microsoft).

Independent of the wider acceptance and use of platforms to support the communication of actors among multiple ecosystems, the coordination of resources shared (e.g., data) is often centrally controlled by an enterprise that is creating value. Users of these platforms trust that the resources they share with others using these platforms are secure and confidential to a certain degree, and recent regulations in Europe and some high-profile abuses (e.g., Facebook/Cambridge Analytica) will continue to draw attention to ways such resource sharing can be made more trustworthy (Davis 2018). In some cases, a third party may be brought in to support communication and the coordination of activities among multiple ecosystems, as in the health information exchanges used to share patient data (Agarwal et al. 2010). However, the success of the health exchanges has been somewhat limited, partially due to a lack of trust in the data shared, limited access to non-clinical participants who may support patient care, or a lack of capacity on the part of some actors to participate in such an exchange (Khuntia et al. 2017).

The potential for blockchain as a platform to distribute and share resources among multiple stakeholders to coordinate activities has attracted the attention of enterprises, including commercial organizations and social institutions in the public and non-profit sectors. Early discourse on its potential and pilot applications led some to jump on the technology bandwagon and seek legitimacy. Such legitimacy is often crucial for enterprises that want to project an image of technology leadership and gain access to resources to compete in the marketplace (Deephouse et al. 2017). To establish a platform as a trusted architecture among actors of multiple ecosystems (Benchoufi and Ravaud 2017; Patel 2019), enabling leadership of ecosystems may position themselves as being innovative and purposefully and strategically allocate resources to explore (or not "miss out" on the opportunities) they can offer (Carroll and Swaminathan 2000).

3.2.1.5 Incentives to Engage Stakeholders in Enterprise Ecosystem

Early adopters of nascent technologies such as blockchain are motivated by their desire to create value that addresses a customer's unmet need (Leblebici et al. 1991; Lounsbury 2001; Sine and Lee 2009; Tolbert and Zucker 1983). The customers of enterprise ecosystems do not have a direct visibility to the technology platform used to create value and do not particularly care, as long as enterprises meet customer expectations and are viewed as taking responsibility for the resource the enterprise collects and distributes.

While value creation is the primary driver or motivator for pioneering early adopters of any technology at the beginning of the adoption cycle (Ansari et al. 2016; Tolbert and Zucker 1983), it is often the uncertainty regarding how the technology can help create value (Resmini and Rosati 2011; Sperber and Wilson 1985) that contributes to challenges for adoption. This is especially true in the case of blockchain technology, which requires that several enterprise stakeholders share resources. The concept of distributed technology with no central coordination is a hard sell to many people who are used to a central entity coordinating such resource sharing. This means that communicating the relevance of the technology as a platform to encourage stakeholder adoption becomes critical.

Sperber and Wilson (1985) suggest that human cognitive processes can be influenced by enhancing and broadening an individual's global perspective. While blockchain development can be application agnostic, multiple examples from different domains where blockchain has addressed key stakeholder needs may help broaden the user's view on the value that the platform provides. Examples here include service coordination of multiple city departments, care delivery for data sharing among multiple health care organizations, and other applications (e.g., credential validation claims or loan processing) where data from multiple agencies are brought together for reconciliation and decision making. In all these examples, value creation needs the trust of many stakeholders outside a single enterprise ecosystem to share resources towards a shared goal.

In general, use of architectures to support communication and sharing of resources has become highly visible and accepted by user communities, ever since the introduction of the Internet/web. However, architectures such as blockchain are asking for the sharing of resources with no central coordination of, at times, personal information of customers, and this may make some of the enterprise stakeholders cautious about its use. Resmini and Rosati (2011) consider that pervasive information architecture should support stakeholder access to information using multiple channels. Given the diversity of stakeholders sharing resources in different application contexts, multimodal social semiotic analysis may be needed to ensure different messages are used to illustrate the value of the technology to these stakeholders (Kress and Van Leeuwen 2001). This means that incentives designed to support stakeholder acceptance of technology have to consider the context within which the applications are being used (e.g., patient privacy in healthcare, currency exchange integrity in financial applications) as well as the messaging used to share the value the technology has to offer. In addition, the setting within which the value of the incentives is perceived is often not the same as the setting that created the technology. It is thus necessary to assess the setting or ecosystem within which stakeholders perceive the value of sharing resources using blockchain technology before developing incentives to influence behavioural change. Diffusion of technology is always based on knowing the adopting stakeholders' characteristics and developing strategies to enable the adopter to evaluate the technology through his own lens. We will come back to this when we discuss trade-offs.

3.2.1.6 Business Model Exploration

The role of a business model is to make the benefits realized from the adoption of a technology artefact outweigh the costs. By sharing tangible resources with stakeholders within their ecosystem using blockchain technology, they are implicitly delegating their trust to the ability of the platform ecosystem to share such resources in a secure manner. Given the impact blockchain architecture has on enterprise features, such as work practices, status hierarchy, and distribution of power relationships, incentives provided to stakeholders have to move beyond operational or financial considerations and address human motivations, including material, psychological, and social dimensions. Within this context, it is important to make the platform adapt to some of the enterprise ecosystem's environment by customizing some of the architectural features to reflect enterprise context.

If the enterprise seeking to use the blockchain platform is a start-up, there needs to be an ecosystem-level effort to reduce the uncertainty it faces and explore channels for sustained funding. For established organizations, incentives may come from the enterprise's need to compete. Incentives to explore some of these new technologies include creating a public impression that the organization is innovative and forward-thinking. Other incentives include demonstrating the pragmatic value the blockchain platform can provide to enhance security, provenance, and immutability of data, which we discuss in more detail below. Often, a few individuals who are particularly passionate about blockchain technology may be incentivized, or enterprises may task the exploration of the technology under its enabling leadership process.

While blockchain technology is presented as a platform that is application agnostic, independent of the enterprise ecosystem, evolving changes in technology as well as enterprise ecosystem dynamics can influence the way business models are used to gain utility from blockchain use. Both technology evolution occurring along each of these three dimensions (architecture, data/record, and social application layers) and the market dynamics of the enterprise system can influence the business models used.

For a blockchain technology to compete as a platform of choice or to be viewed in the mix of platforms that an enterprise considers when addressing resource sharing, it may need to demonstrate effectiveness along a number of dimensions, such as *governance, decentralization, provenance,* and *security features.*

The *governance* defines how enterprises structure their organization and make decisions to reflect the market dynamic within which they operate. The application chosen within this enterprise implicitly reflects the way resources are shared among actors across ecosystems—what resources are shared, and which actors are involved in sharing and using the resources and for how long. The blockchain governance that defines how resource sharing is supported and embedded in the technology has to reflect the social ecosystem (legal, political, or economic environment) within which an enterprise operates, and possibly adapt as the social ecosystem changes.

While *decentralization*, that is, the distribution of resources across multiple independent social actors or enterprises, is designed to bring transparency in decision making and support shared goals and collective action, it can also influence the power dynamics that affect actor behaviour, which is often guided by institutional norms and practices. Therefore, it may be necessary to carefully assess what resources to centralize and what resources to distribute using blockchain. For example, some patient data may be centralized for consistent update by providers and only a link to this data may be stored and shared with appropriate stakeholders using blockchain. On the other hand, patient test data may be stored on the blockchain as it is continually revised by multiple providers when a patient visits them. In other words, the need for consistent and secure sharing of most recent data may dictate from where actors gain access to the data.

Data provenance has been considered critical to understand the origins of data, whether it is the source of raw material used in a product or source of authorship of an idea or artwork, if one wants to trace the source of a product defect or plagiarism or fraud. A timestamp on data when it enters a system is often used to trace the source of product defects, and characteristics of authorship (style, use of certain terms, form of their usage, etc.) may be used to detect fraud in authorship. While *provenance* is key to support auditability of transactions and analysis of temporal data for decision making, the level of precision needed and the degree of completeness that is adequate to realize value can vary. For example, a blockchain application that tracks patient behaviour on smoking cessation counselling may require a different level of data precision to track the impact of counselling methods on patient adherence than tracking cardiac patients at a nursing home to decide on medical interventions.

While the *security* provided to resources exchanged among actors is the key tenet of blockchain, it also should reflect the degree of security needs relevant to the application being supported. For example, a healthcare application that shares patient data may need a different level of security compared to an application that is designed to evaluate the credentials of an applicant applying for a resident job in a hospital.

3.2.1.7 Overcoming Resistance

Resistance to technology adoption, including of blockchain, can occur for many reasons. The enterprise may resist as the application is not necessarily part of the core business or may be viewed as not adding sufficient value. The enterprise may also resist the idea of resource decentralization, if most of its enterprise applications use a centrally-coordinated network to run its operations. Many techniques that are traditionally used to encourage stakeholder adoption, such as providing incentives for the time spent to learn and adopt/use, giving visibility for those who participate, etc. apply here as well.

3.3 Incentives and the Business Model of Blockchain Developer/Administrator Ecosystem

3.3.1 Designing Incentive Mechanisms

In any system where multiple agents interact, the agents may have the ability to take actions that affect their outcomes and the outcomes of other agents. If we assume that agents are self-interested and therefore have an incentive to take actions that will maximize their individual well-being, this may come at the expense of the well-being of other agents.

Situations like this are common. In a supply chain, for example, firms need to interact with suppliers, customers, regulators, and competitors. Each participant in a supply chain has the incentive to take self-interested actions that may lead to a sub-optimal outcome for the supply chain as a whole. For example, a supplier may cut corners on quality to increase its margins and this may lead to the buyer (and customers further down) facing negative consequences.

In settings like this, the design of appropriate rules can alleviate the problem, which in blockchains is performed by the developer ecosystem. Mechanism design, which some authors refer to as the "science of rulemaking" (Hartline and Kleinberg 2012), deals with setting rules for the interaction of independent, interacting, agents. If the rules are set appropriately, it is possible to achieve desirable outcomes. When the rules are not set appropriately, the outcome can be negative.

Hartline and Kleinberg (2012) use the story of the women's doubles badminton tournament in the 2012 summer Olympics in London as an illustrative example of poorly-designed rules that led to a negative outcome. The tournament design (i.e., the "mechanism") involved four groups of four teams each. The first phase of the tournament had a "round robin" format, where the teams in each group played every other team in the group once. The top two teams in each group advanced to the second phase of the tournament, which was a knockout phase. Due to an upset in the round robin phase, a strong team was set to finish in the second place in its group. To avoid meeting this team in the first round of the knockout phase, teams in another group had an incentive to finish second in their group. This led to a farcical situation, where both teams in a match were playing to lose. While this was completely consistent with the incentives of each team, it was a bad outcome for the spectators and the sport.

In any decentralized system, where multiple agents interact to accomplish certain goals, it is important to consider incentives of the individual agents. If the agents face the right incentives, then the performance of the system as a whole can be beneficial to the participants. If not, then the outcomes can be negative or, at a minimum, the system will fail to achieve its objectives. In particular, a key element of a well-designed mechanism is "incentive compatibility" which guarantees that even when agents behave in a self-interested manner, they will make decisions that are beneficial for the entire system.

Any blockchain-based system faces the mechanism design challenge described above. The canonical application, Bitcoin, has an incentive compatible mechanism designed into the protocol. In 2008, Satoshi Nakamoto invented Bitcoin, which uses the Nakamoto Consensus (NC) to realize a public, immutable, and distributed ledger (Nakamoto 2008). The NC protocol has two important components. The first is the so-called Proof-of-Work (PoW) algorithm (Tschorsch and Scheuermann 2016), in which participants (often referred to as miners) are allowed to generate new blocks after successfully solving math puzzles involving hash functions. The second is known as the longest chain rule, by which miners always generate new blocks on top of the longest chain (among competing chains). Although remarkably simple, NC has been formally proven to satisfy blockchain safety and liveness properties as long as a majority of the computing power is controlled by honest miners who strictly follow the NC protocol (Garay et al. 2015, 2017; Kiffer et al. 2018; Pass et al. 2017). The key innovation in Bitcoin was the use of cryptographic techniques and the Proof-of-Work consensus mechanism to ensure that individual agents do not benefit from misrepresenting or tampering with the information recorded in the ledger. In particular, the Bitcoin protocol allows all users to achieve a consensus that the shared and distributed ledger can be trusted.

3.3.2 Bitcoin Blockchain Incentive Challenges

The Bitcoin protocol is an innovative application of mechanism design to achieve consensus in a distributed system; however, the mechanism designed for Bitcoin also has faced several challenges. For example, it is nontrivial to encourage miners to participate in an NC-based blockchain, because miners have to pay for the computing hardware (e.g., CPU, GPU, or ASIC), electricity, and other fees. To address this issue, Nakamoto introduced incentives to NC, by which miners can receive a block reward (i.e., some amount of self-issued tokens) for every block (eventually) included into the longest chain. In addition, miners can also receive transaction fees for all the transactions contained in the block (Nakamoto 2008). These rewards are designed to incentivize miners to contribute their computation power as much as possible. The more computation power a miner contributes, the better chance she is able to solve PoW puzzles.

The incentive mechanism proposed by Nakamoto makes an implicit assumption that all the miners are individually rational (Gervais et al. 2014; Wang et al. 2019). Therefore, a good design should ensure incentive compatibility, which says that miners will suffer from economic loss whenever they deviate from the protocol. Does Bitcoin's incentive mechanism enjoy incentive compatibility? Unfortunately, it does not. Changing the original protocol or code requires consensus from the participants in the network and there is no clear agreement on how to achieve consensus on these kinds of changes. In other words, while the Bitcoin mechanism provides an incentive compatible framework for agents to carry out the normal transactions on the network, there is no mechanism that allows for the necessary periodic updates to the protocol itself.

We now delve more deeply into three incentive mechanism design challenges found in the major permissionless blockchain, Bitcoin and Ethereum.

3.3.2.1 Selfish Mining

As shown in several articles (Eyal and Sirer 2018; Gervais et al. 2016; Nayak et al. 2016; Sapirshtein et al. 2017), if a set of colluding miners deviate from the protocol to maximize their own economic profit, they may obtain a revenue larger than their fair share. Such behaviour is called selfish mining. Specifically, the selfish miners keep their newly mined blocks private and then publish them strategically in order to obtain a higher revenue. By contrast, the honest miners immediately publish their newly mined blocks. To see how it works, imagine that the selfish miners already have two blocks in private while the honest miners still mine on the public chain (which is two blocks shorter than the private chain). When some honest miner mines a new block, the selfish miners will publish these two private blocks immediately. According to the longest chain rule, all the honest miners will accept these two blocks and reject the block mined by the honest miner. In this case, the selfish miners not only receive two block rewards (as well as the associated transaction fees), but also make the honest block useless since it is no longer in the longest chain.

The selfish mining attack was first proposed in the Bitcoin forum. Later on, Eyal and Sirer (2018) developed a Markov model to analyze a particular selfish mining strategy in. They showed that the threshold of the computational power to make selfish mining profitable is 25% under the longest chain rule with uniform tie-breaking policy. Inspired by their work, Nayak et al. (2016) expanded the mining spaces and introduced a new mining strategy, which leads to higher revenue for selfish miners. They also considered the role of eclipse attacks on selfish mining.

Sapirshtein et al. (2017) used a Markov Decision Process (MDP) to generalize various selfish mining strategies and demonstrated that the optimal strategy has a threshold of 23.2%. Furthermore, the effect of network delay was considered for selfish mining in Eyal and Sirer (2018) and Sapirshtein et al. (2017).

In Bitcoin, miners receive two types of mining rewards: block rewards and transaction fees. Since the block reward is the dominant reward in today's Bitcoin, the previous studies often ignore the impact of transaction fees. As Bitcoin's block reward halves every 4 years on average, transaction fees will eventually play a critical role. Motivated by this, a Bitcoin-like system without block rewards is considered in (Carlsten 2016; Carlsten et al. 2016), which shows that Bitcoin mining is no longer stable in that miners will create forks on purpose to steal the high transaction fees in some "wealthy" blocks. In this case, the threshold of making selfish mining profitable will be arbitrarily low.

Clearly, different mining reward settings lead to different optimized selfish mining strategies. For instance, if we turn to Ethereum (today's biggest decentralized platform that runs smart contracts) we will find four types of mining rewards: block reward, uncle block reward, nephew block reward, and gas cost (Wood 2018). Here, gas cost is similar to transaction fees. Such a reward structure makes the analysis of selfish mining more complicated. Gervais et al. (2016) developed a quantitative framework to analyze selfish mining in various blockchains (including Ethereum and Bitcoin). Ritz and Zugenmaier (2018) built a Monte Carlo simulation platform to quantify the impact of selfish mining in Ethereum. Niu and Feng (2019) introduced a two-dimensional Markov process to model the behaviour of a particular selfish mining strategy in Ethereum and found that the uncle and nephew rewards can lower the threshold of computation power. In addition, they studied the impact of different uncle reward functions.

As shown in the above studies, selfish mining strategies result in a waste of computing power by encouraging forks on purpose, which in turn undermines the security of a blockchain system. This motivates numerous defense mechanisms. As noted in Niu and Feng (2019), Heilman proposed a defense mechanism called Freshness Preferred, which uses the latest un-forgeable timestamp issued by a trusted party (Heilman 2014). With this mechanism, the threshold of selfish mining can be increased to 32%. Bahack (2013) introduced a fork-punishment rule to make selfish mining unprofitable. To do this, each miner in the system can include a fork evidence in their block. Once confirmed, the miner can get half of the total rewards of the winning branch. Zhang and Preneel (2017) proposed a new fork-resolving policy called weighted FRP, in which forks are resolved by comparing all chains' weights, and the hidden blocks mined by the selfish miner cannot contributes any weights to its own branch. Thus, the leading block advantage of the selfish miner can be reduced. Pass and Shi (2017) also proposed a fair blockchain protocol called Fruitchains, in which rewards and transaction fees are evenly distributed among the miners and no selfish mining can be made profitable.

3.3.2.2 Recentralization/Pool Mining

In order to reduce the variance of miners' revenue, individual miners can form mining pools to mine blocks together and share the revenue according to individuals' computation power. Nowadays the top six mining pools in Bitcoin and Ethereum have occupied over 70% of the total computation power (BTC.com 2019; Etherscan 2019). Clearly, the presence of the mining pools hurts the blockchain's decentralization, making the system vulnerable. In addition, the reward distributing policy used by mining pool leads to new incentive issues.

In Luu et al. (2015) and Tosh et al. (2017), the authors proposed a block withholding (BWH) mining strategy, in which the selfish miner in the pool can increase their own revenue while decreasing honest miners' share in the pool. Here the mining pools are assumed to adopt a pay-per-share (PPS) protocol (Rosenfeld 2011). When adopting this mining strategy, the selfish miner will split their computation power into several different pools: one pool with most of the computation power used for honest mining and others for launching BWH. In the victim pools, the selfish miner submits all work shares except the valid puzzle solutions. It's obvious that the selfish miner will suffer some economic loss in the victim pools. But the selfish miner can obtain more from the honest mining. Indeed, it is shown that the selfish miner can always gain more revenues by mining dishonestly regardless of its computation power. In addition, it is shown that some big mining pools can dominate the network through BWH attacks on smaller mining pools.

Luu et al. (2015) and Tosh et al. (2017) assumed that there is only one selfish miner adopting BWH, and all the other miners behave honestly. In Eyal (2015), the authors considered a more complicated case where mining pools attack each other with BWH. They found that no matter how many pools exist in the network, no-pool-attack is not a Nash equilibrium (i.e., mining pools tend to attack each other for their own benefits). Additionally, if mining pools attack each other, every pool will earn less than they would have if none had attacked. Thus, for two pools, the decision whether or not to attack is the miner's dilemma. Based on these works, Bag et al. (2017) found that all mining pools except the victim mining pools will benefit from the BWH attack even if they do not launch the BWH attack. This finding implies that these pools will not report the BWH behaviours unless they are the victims, and they even may sponsor the attacker to launch the BWH attack. Inspired by the BWH attack, Kwon et al. (2017) proposed a new attack called fork-after-withholding (FAW). FAW is shown to be always equally or more profitable than BWH attack. Furthermore, when two pools execute FAW attacks on each other, the miner's dilemma may not hold: under certain circumstances, the larger pool can consistently win.

Apart from BWH attacks, selfish miners can also hop between different mining pools and utilize the reward distribution policy to gain more profits. For example, in the pay-per-share protocol, payments are calculated based on rounds (Rosenfeld 2011), which are the time intervals between mined blocks in series. The longer the round, the more shares submitted and the less each share is worth. This implies that a share submitted early will gain a higher reward. Thus, the selfish miners prefer to mine in the pool at the beginning of a round and hop to other pools when the round is too long. Rosenfeld (2011) showed that when the accumulated shares of the mining pool in a round exceed a threshold, it will be more profitable for miners to hop to other pools or mine by themselves. Based on these observations, some protocols are proposed to address the mining pool hopping problem. Slushpool (2019) implemented a score-based method, in which shares are weighted. The shares submitted at the beginning of a round get more scores, while later submitted shares gain less scores. Rosenfeld (2011) has also analyzed other reward protocols such as pay-per-last-N shares and payment-contract-based.

3.3.2.3 Token Incentives

In the previous sections, we focused on the incentive study of blockchains from a technical perspective, analyzing NC's incentive from the individual miners' and mining pools' perspectives. The previous incentive studies are built on the assumption that the parties involved in mining are economically rational and likely to win the mining rewards (self-issued tokens). In other words, these parties must agree on the tokens' value. Kroll et al. (2013) suggested that some attacker can launch a Golden Finger attack to destroy the Bitcoin economy in order to achieve utility outside the Bitcoin economy. Once destroyed, no miners would like to participate in NC for some non-value tokens.

It is easy to see that the higher the value of tokens, the more computational power is involved in the mining and the more secure the system is. In return, a secure system will attract more users, and then raise the token's value, which forms a positive feedback loop. This leads to numerous studies on the economics of blockchain-based tokens and Initial Coin Offerings (see, for example, Feng et al. 2018; Rohr and Wright 2017), which is beyond the scope of this chapter.

3.3.3 Permissioned Blockchains

In the discussions thus far, we focused on permissionless blockchains, where anyone can participate in adding a data block to engage in sharing resources, financial or otherwise, with a peer node on the network. Many commercial enterprises entering into this space view "proof-of-consensus" or "proof-of-authority" as suitable to gain the value the blockchain platform provides without the computational resources needed to create blocks of data using PoW discussed earlier. While this brings about some of the concepts of central coordination, at least with regard to who can contribute to resource sharing, it still preserves the security of the actual data that is shared as well as who coordinates with whom on activities involved in sharing the data. With recent examples of Walmart and IBM using blockchain to track problems in the food supply chain (Nasdaq.com 2017), health care providers providing patients with access to their data so they can share with other providers on the network to get second opinions (Embleema 2019) and streamline selected data sharing, and banks like Chase and social media companies like Facebook working with partners to support peer-to-peer financial transactions and exchange of personal information (Allison 2017; CB Insights 2019), permissioned blockchains are becoming viable platforms to allow communication and coordination of ecosystem user activities.

In these cases, the incentive for blockchain stakeholders shifts from miners who create data blocks to the entire team of platform administrators and data/record block coders (as a team). The business model for the team such as IBM, unless it is a part of an enterprise's exploratory team, is to develop the blockchain platform that pays for the resources to store and transact resources (e.g. Ether or other crypto currency) and attracts ecosystem users like Walmart to pay for the value that such a platform provides. Such co-creation of value with customers or suppliers is becoming a common practice to sustain competitiveness in the digital age.

3.4 Trade-offs

Actor-network theory (Latour 1999) argues that machines and humans, as they interact towards accomplishing a goal, generate sustained value creation. Blockchain technology, acting as a platform to support communication and coordination of stakeholders (systems as well as human actors), must adapt to the changing dynamic in the digital age of both technology and other environmental and customer dynamics. To this end, some trade-offs must be made across ecosystems as discussed next.

From the perspective of data/records management, the main incentive for blockchain adoption is that it can improve the privacy and security of records (Patel 2019). However, realizing business objectives, such as value creation, resource acquisition, and legitimacy, may be in conflict with this main incentive.

First, when businesses pursue the practical value of Blockchain, realizing blockchain's potential to facilitate data sharing, they may inevitably face the risk of hampering privacy. For example, when some organizations initiated a consortium to share data on blockchain by combining previously siloed data for research and innovation (Wilson 2019; Yafimava 2019), this immediately brought up the issue of users' private information being shared with entities that they have not consented to, putting their privacy at risk. Similarly, when organizations try to use blockchain to improve supply chain management (Grodal 2018) and stamp the identity of suppliers on the final product, it leads to exposing key information about suppliers and potentially eroding their privacy. Overall, as much of blockchain's practical value resides in its potential to make records transparent and shareable, pursuing this incentive—the practical value of blockchain—may inherently jeopardize privacy.

Second, as entrepreneurial businesses adopt blockchain for the sake of the resources they can gain from its adoption, they may create pre-mature applications that do not have a fully-developed system to protect data security. Also, the competition for resources can be intense when established businesses try to adopt blockchain for the sake of winning recognition or resource-based advantages (Grodal 2018). They may fail to fully investigate the technical landscape of blockchain and not be careful and strategic in selecting blockchain developers. Given the need to gain access to many users to make the blockchain platform relevant, enterprises may not carefully design the security system and may consume a good part of the resource to deliver on the promises their solution aims to provide. A few incidences have happened where early developers of a blockchain-based currency system or platform were attacked by unknown sources, and a large amount of user information was exposed (e.g., the hacks to Bitthumb Exchange and Coinrail). In the infamous case of Quadriga, individually owned cryptocurrency was appropriated, suggesting the possibility that, early on, a blockchain system may give space for a few individuals to take advantage of others' data (Alexander 2019). In summary, early adopters may be incentivized by the resources associated with blockchain, but such incentives can lead them to focus less on data security, potentially causing security problems.

Third, when businesses adopt blockchain for the sake of legitimacy, it may lead to symbolic or ceremonial adoption, which can put both data security and privacy at risk. Organizational research has well-established theories on the issue of legitimacy when new technologies are adopted, and organizations use a number of tactics to protect their business core from being disrupted by the adoption (Meyer and Rowan 1977; Scott 2014): These tactics include decoupling, "the logic of confidence", and impression management (Scott 2014). Organizations are very likely to deploy such tactics when the technology is under-developed or cannot be well-integrated into the technical core.

In the case of blockchain, organizations may perceive that the adoption of blockchain is a signal to stakeholders that they are becoming sophisticated in the use of innovative technologies, embracing innovations, and actively exploring new ways to create value for customers. As such, they may focus on marketing blockchain adoption rather than substantively exploring how it can be leveraged to create sustained value and potentially be embedded within the business practice. While rushing to show value, they may implement incomplete systems, not invest the needed resources in early-stage application development, and pay limited attention to the concerns of those interested in addressing data security and privacy. For these reasons, the effective use of enabling leadership—to explore technologies such as blockchain with limited scope and deliberate reflection and decision making on when, what, and how much to adopt by adaptive and administrative leadership—is critical.

In summary, business incentives to help enterprises adopt blockchain to meet shared goals of several actors across ecosystems have to weigh the technology tradeoffs, especially those with respect to data/records management standards used to address data security and privacy. The challenges here may be addressed by knowing what data or resources to share among actors, who among the ecosystem actors should have access to the data, and the scope of blockchain platform use, as not all ecosystem system communication and coordination needs to occur through a single platform. In fact, in some healthcare domains, resources such as patient data may be shared using a centrally coordinated data communication architecture (e.g., those in a hospital network), with other resources shared via the blockchain platform (e.g., non-clinical care provider sharing of discharge plan activities).

3.5 Conclusions and Directions for Future Research

3.5.1 Conclusions

The goal of this chapter is to find a theoretical framework for understanding the role of incentives in the development and use of blockchain technology. The earlier sections compared incentives and motivation, and developed the role of context in making change relevant and in influencing behaviour. Given that a change in behaviour is needed among multiple stakeholders and that the gaining of their trust is important for the successful use of a distributed resource sharing platform such as blockchain, this chapter discussed the role of incentives for stakeholders in each of the two distinct ecosystems: developers, and users. Categorizing blockchain architecture as a platform that connects three layers—technology, data/records, and social/application—both incentives and business models were discussed, as they influence the behaviour of miners in adding data/records to support resource sharing, and users in adopting the technology to share resources. With the evolving technology and business landscape in today's complex market dynamic, the incentives and business models used to influence the stakeholders of each ecosystem must be adapted to reflect changes in the other ecosystem.

A key takeaway from the above observations is that any blockchain-based system is not just a technical system, but also a socio-informational-technical one. In other words, as Werbach (2018) points out, the rule for the continued successful operation of such systems should be not be based only on dry (technological) code but should also consider wet code (i.e., any mechanisms that operate and are enforced outside of the dry, i.e., human, code), and we would argue, information as well.

In practice, we can interpret this to mean that incentive design in blockchain-based systems requires a careful combination of innovative technological solutions and more traditional information and social governance mechanisms. While these issues are relevant to cryptocurrencies like Bitcoin and others, they are even more relevant in the design and operation of permissioned blockchains. In a permissioned blockchain, unlike in a purely peer-to-peer system, a participant in the network often takes a leadership position and is interested in getting other agents to participate in the network.

For example, a supply chain application like tracking the provenance of food may involve a large retailer like Walmart who initiates a permissioned blockchain and requires the participation of suppliers and other stakeholders. In this setting, Walmart would need to ensure that suppliers and others have the right incentives (to share information truthfully, etc.). Problems of this type are well studied in the principal-agent literature, where a "principal" (e.g. Walmart) would take on the role of a mechanism designer and set the rules for the other agents to follow.

Incentive design in blockchain-based systems is an emerging area of research and application. While mechanism design theory and principal-agent theory are well developed, the design of incentives in blockchain-based systems offers new and interesting challenges. Firstly, the use of cryptographic methods to achieve incentive-compatibility is an underexplored topic. The success of Bitcoin and other cryptocurrencies is a testament to the power of this approach. But, as noted earlier, there are several challenges (like the need for significant computing requirements) that may limit the ability to scale this approach. Secondly, for blockchain-based systems to achieve widespread adoption and success, incentive designers need to recognize that other, more traditional, governance structures must complement the more recent technological innovations in the design of incentives.

3.5.2 Future Research

Future research calls for a focus on how effectively blockchain technology and the application domain adapt to each other in an evolving ecosystem dynamic. One way to address adaptation to changes in a customer ecosystem is to build agility in business operations. In the case of blockchain technology, it must adapt to the volatility of the application domain. For example, for applications such as finance and real estate, the type of resource shared and the legal binding contracts that are to be enforced with trust are relatively stable. Therefore, as the blockchain technology continues to evolve, the technology and data/record layers must adapt to application requirements, and the incentives must support the need to ensure the integrity and auditability of the transactions. On the other hand, in applications such as healthcare, where the nature of the resource shared and the regulatory and user behaviours continue to evolve, technology platform needs to support agility in the way data/records are structured. As discussed earlier, some patient data may be stored centrally with only links to the data made available on the blockchain, and other patient data like test results may be shared via blockchain because of the need to gain access to the most recent data for diagnosis. Also, the size of data may dictate where it is stored, as in the case of a patient's radiology scans. Lastly, when the scale and scope of blockchain applications are hard to justify in their current form, but if the technology does offer potential benefits over time, agility may be needed to allow a mix of options for exploration: permissioned blockchain to begin with, and then a transition to a more hybrid model as viability becomes well established (e.g., mix of centralized, permissioned blockchain, or permissionless blockchain).

On the other hand, information systems theory argues that resources shared using blockchain architecture may need to be tailored to address the needs of various application user groups (Orlandi et al. 2018). Even though this research has suggested a layered approach to architect resource sharing among high-performance individuals, the approach may be tailored to meet the needs of diverse user populations if the resources shared are tailored to the roles users play within an organization and/or their technological capabilities. For example, in healthcare applications, some users are physicians and care providers who share patient data, while others are social workers that support patients, who need to engage in activities to adhere to care plans. The layered approach illustrated briefly below potentially illustrates how such adaptation could be achieved.

- **Knowledge**—share learning objects using metadata, concepts of classification and ontology
  - Disease classifications and reimbursement codes; naming of health conditions or symptoms
- **Information models**—used to structure the shared information (institutional environment, labeling of information products, design architectures, etc.)
  - Regulations and reimbursement policies and new medical devices; dosage information on drugs; caloric content of foods
- **Relevance**—learning objects relevant to high performance professionals (Resmini and Rosati 2011)
  - Certifications and quality guidelines for physicians and nurses; infection control practices and immunization guidelines
- **Multimodality**—multimodal learning objects incorporated as didactic material (Kress and Van Leeuwen 2001)
  - Research reports and clinical practice guidelines; use case scenarios and simplified guidelines on public health practices
- **Gamification**—learning games as a means of increasing user engagement (Poole et al. 2014).
  - Simulation games to test alternate treatment scenarios; games to engage seniors or patients to complete a set of activities post-discharge
- **User experience**—influence concepts to improve user experience
  - Chat sessions with experts on clinical diagnosis consultations; chats with peers or peer rankings on adherence practices (Nielsen 1999)

In summary, future research is needed to determine the focus of agility among the technology, data/record, and social/application layers to ensure that they can adapt to the dynamics of the ecosystems and its actor characteristics. Ultimately, the relevance of the platform, as it addresses the intrinsic or extrinsic drivers of the stakeholders within each ecosystem, operating and interacting within a broader blockchain ecosystem, can act as a powerful incentive to successfully assimilate blockchain technology.

References

Agarwal, R., Crowley, P. K., Khuntia, J., & Mithas, S. (2010). *The District of Columbia regional health information organization (DC RHIO): Current progress and the road ahead*. College Park, MD: Center for Health Information and Decision Systems, Robert H. Smith School of Business, University of Maryland. Retrieved from https://www.rhsmith.umd.edu/files/Documents/Centers/CHIDS/DCRHIOAssessmentReport.pdf

Aghena, W., De Smet, A., & Weerda, K. (2015). Agility: It rhymes with stability. *McKinsey Quarterly, 2015*(4), 2–9. Retrieved from https://www.mckinsey.com/business-functions/organization/our-insights/agility-it-rhymes-with-stability

Alexander, D. (2019). Quadriga founder transferred clients' cryptocurrency to his own personal accounts, Ernst & Young finds. *The Financial Post*. Retrieved November 9, 2019, from https://business.financialpost.com/technology/blockchain/quadriga-founder-transferred-clients-cryptocurrency-to-his-own-personal-accounts-ernst-young-finds

Allison, I. (2017). Quorum: J.P. Morgan's Ethereum fork could eat your lunch. *International Business Times*. Retrieved November 14, from https://www.ibtimes.co.uk/quorum-j-p-morgans-ethereum-fork-could-eat-your-lunch-1625606

Ansari, S., Garud, R., & Kumaraswamy, A. (2016). The disrupter's dilemma: TiVo and the U.S. television ecosystem. *Strategic Management Journal, 37*(9), 1829–1853. https://doi.org/10.1002/smj.2442.

Bag, S., Ruj, S., & Sakurai, K. (2017). Bitcoin block withholding attack: Analysis and mitigation. *IEEE Transactions on Information Forensics and Security, 12*(8), 1967–1978. Retrieved from http://ieeexplore.ieee.org/stamp/stamp.jsp?tp=&arnumber=7728010&isnumber=7921660

Bahack, L. (2013). Theoretical Bitcoin attacks with less than half of the computational power (draft). *ArXiv Preprint*. arXiv:1312.7013

Benchoufi, M., & Ravaud, P. (2017). Blockchain technology for improving clinical research quality. *Trials, 18*, 335–340. https://doi.org/10.1186/s13063-017-2035-z.

Bossert, O., Laartz, J., & Ramsey, T. J. (2014). Running your company at two speeds. *McKinsey Quarterly, 2014*(4), 12–14. Retrieved from https://www.mckinsey.com/business-functions/mckinsey-digital/our-insights/running-your-company-at-two-speeds

Brandao, O. C., & Duque, C. G. (2011). Comunicação científica contemporânea e de vanguarda. In O. C. Brandao & C. G. Duque (Eds.), *Ciência da Informação Estudos e Práticas* (pp. 9–33). Brasília: Centro Editorial.

BTC.com. (2019). *Top miners by blocks in Bitcoin*. Retrieved January 24, 2020, from https://btc.com/stats/pool

Carlsten, M. (2016). *The impact of transaction fees on Bitcoin mining strategies* (Doctoral dissertation). Retrieved from https://www.cs.princeton.edu/research/techreps/TR-983-16

Carlsten, M., Kalodner, H., Weinberg, S. M., & Narayanan, A. (2016). On the instability of Bitcoin without the block reward. In *CCS'16: Proceedings of the 2016 ACM SIGSAC Conference on Computer and Communications Security* (pp. 154–167). New York: ACM. https://doi.org/10.1145/2976749.2978408.

Carroll, G. R., & Swaminathan, A. (2000). Why the microbrewery movement? Organizational dynamics of resource partitioning in the U.S. brewing industry. *American Journal of Sociology, 106*(3), 715–762. https://doi.org/10.1086/318962.

CB Insights. (2019). FaceCoin: Here's what Facebook could build in blockchain and cryptocurrency. *CBInsights.com*. Retrieved November 26, 2019, from https://www.cbinsights.com/research/facebook-blockchain-cryptocurrency/

Davis, R. (2018). *What the Cambridge Analytica revelations signal for future political campaigns*. Retrieved February 27, 2020, from https://observer.com/2018/03/cambridge-analytica-revelations-signal-future-political-campaigns

Deephouse, D. L., Bundy, J., Tost, L. P., & Suchman, M. C. (2017). Organizational legitimacy: Six key questions. In R. Greenwood, C. Oliver, T. Lawrence, & R. Meyer (Eds.), *The SAGE handbook of organizational institutionalism* (2nd ed., pp. 27–70). Thousand Oaks, CA: Sage.

Eisenmann, T. R., Parker, G., & Van Alstyne, M. (2009). Opening platforms: How, when and why? In A. Gawer (Ed.), *Platforms, markets and innovation* (pp. 131–162). Cheltenham: Edward Elgar.

Embleema. (2019). *Home*. Retrieved November 15, 2019, from https://www.embleema.com/

Etherscan. (2019). *Top 25 miners by blocks in Ethereum*. Retrieved January 24 2020, from https://etherscan.io/stat/miner?blocktype=blocks

Eyal, I. (2015). The miner's dilemma. In *SP'15: Proceedings of the 2015 IEEE Symposium on Security and Privacy* (pp. 89–103). San Jose, CA: IEEE Computer Society. https://doi.org/10.1109/SP.2015.13.

Eyal, I., & Sirer, E. G. (2018). Majority is not enough: Bitcoin mining is vulnerable. *Communications of the ACM, 61*(7), 95–102. https://doi.org/10.1145/3212998.

Feng, C., Li, N., Lu, B., Wong, M., & Zhang, M. (2018). Initial coin offerings, blockchain technology, and white paper disclosures. *SSRN*. https://doi.org/10.2139/ssrn.3256289.

Garay, J. A., Kiayias, A., & Leonardos, N. (2015). The Bitcoin backbone protocol: Analysis and applications. In E. Oswald & M. Fischlin (Eds.), *Advances in cryptology – EUROCRYPT 2015. Lecture notes in computer science* (Vol. 9057, pp. 281–310). Berlin: Springer. https://doi.org/10.1007/978-3-662-46803-6_10.

Garay, J. A., Kiayias, A., & Leonardos, N. (2017). The Bitcoin backbone protocol with chains of variable difficulty. In J. Katz & H. Shacham (Eds.), *Advances in cryptology – CRYPTO 2017. Lecture notes in computer science* (Vol. 10401, pp. 291–323). Cham: Springer International. https://doi.org/10.1007/978-3-319-63688-7_10.

Gerst, M. H., & Bunduchi, R. (2006). Intraorganisational power and the adoption of interorganisational IT innovations: The inside story of Covisint. *International Journal of Technology Intelligence and Planning, 3*(1), 57–74. https://doi.org/10.1504/IJTIP.2007.013038.

Gervais, A., Karame, G. O., Capkun, V., & Capkun, S. (2014). Is Bitcoin a decentralized currency? *IEEE Security Privacy, 12*(3), 54–60. https://doi.org/10.1109/MSP.2014.49.

Gervais, A., Karame, G. O., Wüst, K., Glykantzis, V., Ritzdorf, H., & Capkun, S. (2016) On the security and performance of proof of work blockchains. In *CCS'16: Proceedings of the 2016 ACM SIGSAC Conference on Computer and Communications Security* (pp. 3–16). Vienna: ACM. https://doi.org/10.1145/2976749.2978341.

Grodal, S. (2018). Field expansion and contraction: How communities shape social and symbolic boundaries. *Administrative Science Quarterly, 63*(4), 783–818. https://doi.org/10.1177/F0001839217744555.

Hartline, J., & Kleinberg, R. (2012). Badminton and the science of rule making. *Huffington Post*. Retrieved from http://jasonhartline.com/HuffingtonPost-2012-badminton.pdf

Heilman, E. (2014). One weird trick to stop selfish miners: Fresh Bitcoins, a solution for the honest miner. In R. Böhme, M. Brenner, T. Moore, & M. Smith (Eds.), *Financial cryptography and data security. FC 2014. Lecture notes in computer science* (Vol. 8438, pp. 161–162). Berlin: Springer.

Incentive. (2015). *APA dictionary of psychology* (2nd ed.). Washington, DC: American Psychological Association.

Khuntia, J., Yim, D., Tanniru, M., & Lim, S. (2017). Patient empowerment and engagement with a health infomediary. *Health Policy and Technology, 6*(1), 40–50. https://doi.org/10.1016/j.hlpt.2016.11.003.

Kiffer, L., Rajaraman, R., & Shelat, A. (2018). A better method to analyze blockchain consistency. In *CCS'18: Proceedings of the 2018 ACM SIGSAC Conference on Computer and Communications Security* (pp. 729–744). New York: ACM. https://doi.org/10.1145/3243734.3243814.

Kress, G. R., & Van Leeuwen, T. (2001). *Multimodal discourse: The modes and media of contemporary communication*. London: Hodder Education.

Kroll, J. A., Davey, I. C., & Felten, E. W. (2013). The economics of Bitcoin mining, or Bitcoin in the presence of adversaries. In *Proceedings of the Twelfth Workshop on the Economics of*

Information Security (WEIS 2013) (pp. 11–32), Washington, DC, June 11–12, 2013. Retrieved from https://www.econinfosec.org/archive/weis2013/papers/KrollDaveyFeltenWEIS2013.pdf

Kwon, Y., Kim, D., Son, Y., Vasserman, E., & Kim, Y. (2017). Be selfish and avoid dilemmas: Fork after withholding (faw) attacks on Bitcoin. In *CCS'17: Proceedings of the 2017 ACM SIGSAC Conference on Computer and Communications Security* (pp. 195–209). New York: ACM.

Latour, B. (1999). On recalling ANT. *The Sociological Review, 47*(Suppl 1), 15–25. https://doi.org/10.1111/j.1467-954X.1999.tb03480.x.

Leblebici, H., Salancik, G. R., Copay, A., & King, T. (1991). Institutional change and the transformation of interorganizational fields: An organizational history of the U.S. radio broadcasting industry. *Administrative Science Quarterly, 36*(3), 333–363. Retrieved from https://www.jstor.org/stable/2393200

Lounsbury, M. (2001). Institutional sources of practice variation: Staffing college and university recycling programs. *Administrative Science Quarterly, 46*(1), 29–56. https://doi.org/10.2307/2667124.

Lusch, R. F., & Nambisan, S. (2015). Service innovation: A service-dominant logic perspective. *MIS Quarterly, 39*(1), 155–175.

Luu, L., Saha, R., Parameshwaran, I., Saxena, P., & Hobor, A. (2015). On power splitting games in distributed computation: The case of Bitcoin pooled mining. In *Proceedings of the 2015 IEEE 28th Computer Security Foundations Symposium* (pp. 397–411). Verona: IEE Computer Society. https://doi.org/10.1109/CSF.2015.34

Marion, R. (2008). Complexity theory for organizations and organizational leadership. In M. Uhl-Bien & R. Marion (Eds.), *Complexity leadership* (pp. 1–15). Charlotte, NC: Information Age.

Meyer, J., & Rowan, B. (1977). Institutionalized organizations: Formal structure as myth and ceremony. *American Journal of Sociology, 83*(2), 340–363. https://doi.org/10.1086/226550.

Nakamoto, S. (2008). *Bitcoin: A peer-to-peer electronic cash system.* Satoshi Nakamoto Institute. Retrieved from https://nakamotoinstitute.org/bitcoin/

nasdaq.com. (2017). *IBM and Walmart launch blockchain-based food safety alliance for China.* Retrieved November 14, 2019, from https://tinyurl.com/ybu9xff7

Nayak, K., Kumar, S., Miller, A., & Shi, E. (2016). Stubborn mining: Generalizing selfish mining and combining with an eclipse attack. In *2016 IEEE European Symposium on Security and Privacy* (EuroS&P) (pp. 305–320). Saarbrucken. Los Alamitos, CA: IEEE Computer Society. https://doi.org/10.1109/EuroSP.2016.32

Nielsen, J. (1999). *Designing web usability: The practice of simplicity.* Thousand Oaks, CA: New Riders.

Niu, J., & Feng, C. (2019). Selfish mining in Ethereum. *ArXiv Preprint.* arXiv:1901.04620

Orlandi, T. R. C., Duque, C. G., Mori, A., & Bernardo, C. G. (2018). *A new model of information architecture associated with multimodality for training high performance professionals.* Poster presented at the 81st Annual Meeting of the American Society for Information Science and Technology, Vancouver, 10–14 November 2018.

Pass, R., & Shi, E. (2017). Fruitchains: A fair blockchain. In *PODC'17: Proceedings of the ACM Symposium on Principles of Distributed Computing* (pp. 315–324). New York: ACM. https://doi.org/10.1145/3087801.3087809.

Pass, R., Seeman, L., & Shelat, A. (2017). Analysis of the blockchain protocol in asynchronous networks. In J. S. Coron & J. B. Nielsen (Eds.), *Advances in cryptology – EUROCRYPT 2017. Lecture notes in computer science* (Vol. 10211, pp. 643–673). Cham: Springer International. https://doi.org/10.1007/978-3-319-56614-6_22.

Patel, V. (2019). A framework for secure and decentralized sharing of medical imaging data via blockchain consensus. *Health Informatics Journal, 25*(4), 1398–1411. https://doi.org/10.1177/1460458218769699.

Poole, S., Kemp, E., Patterson, L., & Williams, K. (2014). Get your head in the game: Using gamification in business education to connect with generation Y. *Journal for Excellence in Business Education, 3*(2), 1–9. http://www.jebejournal.org/index.php/jebe/article/view/40

Resmini, A., & Rosati, L. (2011). *Pervasive information architecture: Designing cross-channel user experiences*. Burlington: Elsevier Science.

Ritz, F., & Zugenmaier, A. (2018). The impact of uncle rewards on selfish mining in Ethereum. In *2018 IEEE European Symposium on Security and Privacy Workshops* (EuroS&PW) (pp. 50–57). London: IEEE. https://doi.org/10.1109/EuroSPW.2018.00013.

Rohr, J., & Wright, A. (2017). Blockchain-based token sales, initial coin offerings, and the democratization of public capital markets. *Cardozo Legal Studies Research Paper*, no. 527. https://doi.org/10.2139/ssrn.3048104.

Rosenfeld, M. (2011). Analysis of Bitcoin pooled mining reward systems. *ArXiv Preprint*. arXiv:1112.4980

Saif, S. K., Nawaz, A., Jan, F. A., & Khan, M. I. (2012). Synthesizing the theories of job-satisfaction across the cultural/attitudinal dimensions. *Interdisciplinary Journal of Contemporary Research in Business, 3*(9), 1382–1396.

Sapirshtein, A., Sompolinsky, Y., & Zohar, A. (2017). Optimal selfish mining strategies in Bitcoin. In J. Grossklags & B. Preneel (Eds.), *Financial cryptography and data security. FC 2016. Lecture notes in computer science* (Vol. 9603, pp. 515–532). Berlin: Springer.

Scott, W. R. (2014). *Institutions and organizations: Ideas, interests, and identities* (4th ed.). Los Angeles: Sage.

Sine, W. D., & Lee, B. H. (2009). Tilting at windmills? The environmental movement and the emergence of the U.S. wind energy sector. *Administrative Science Quarterly, 54*(1), 123–155. https://doi.org/10.2189/asqu.2009.54.1.123.

Slushpool. (2019). *Revenue system*. Retrieved January 24, 2020, from https://slushpool.com/help/reward-system/

Sperber, D., & Wilson, D. (1985). *Relevance: Communication and cognition*. Oxford: Blackwell.

Spohrer, J., Anderson, L., Pass, N., & Ager, T. (2008). *Service science and service-dominant logic*. Otago Forum 2 (2008) – Paper no. 2. https://doi.org/10.1287/serv.1.1.32.

Tolbert, P., & Zucker, L. (1983). Institutional sources of change in the formal structure of organizations: The diffusion of civil service reform, 1880-1935. *Administrative Science Quarterly, 28*(1), 22–39. Retrieved from https://digitalcommons.ilr.cornell.edu/cgi/viewcontent.cgi?article=1132&context=articles

Tosh, D. K., Shetty, S., Liang, X., Kamhoua, C. A., Kwiat, K. A., & Njilla, L. (2017). Security implications of blockchain cloud with analysis of block withholding attack. In *2017 17th IEEE/ACM International Symposium on Cluster, Cloud and Grid Computing (CCGRID),* Madrid (pp. 458–467). https://doi.org/10.1109/CCGRID.2017.111

Tschorsch, F., & Scheuermann, B. (2016). Bitcoin and beyond: A technical survey on decentralized digital currencies. *IEEE Communications Surveys Tutorials, 18*(3), 2084–2123. https://doi.org/10.1109/COMST.2016.2535718.

Uhl-Bien, M., Marion, R., & McKelvey, B. (2007). Complexity leadership theory: Shifting leadership from the industrial age to the knowledge era. *The Leadership Quarterly, 18*(4), 298–318. https://doi.org/10.1016/j.leaqua.2007.04.002.

Vargo, S. L., & Lusch, R. F. (2006). Service-dominant logic: What it is, what it is not, what it might be. In S. L. Vargo & R. F. Lusch (Eds.), *The service-dominant logic of marketing: Dialog, debate, and directions* (pp. 43–55). Armonk, NY: M. E. Sharpe.

Wang, W., Hoang, D. T., Hu, P., Xiong, Z., Niyato, D., Wang, P, Wen, Y, & Kim, D. I. (2019). A survey on consensus mechanisms and mining management in blockchain networks. *ArXiv Preprint*, 1–33. arXiv:1805.02707

Werbach, K. (2018). *The blockchain and the new architecture of trust*. Cambridge, MA: MIT.

Wieland, H., Vargo, S., Akaka, M. A., & Barbeau, B. (2018). A unifying perspective for the technological, business model, and market aspects of innovation. In R. F. Lusch & S. L. Vargo (Eds.), *Sage handbook on service dominant logic* (pp. 508–521). London: Sage.

Wilson, R. (2019). The right way to do Blockchain consortiums. *coindesk*. Retrieved November 9, 2019, from https://www.coindesk.com/the-right-way-to-do-blockchain-consortiums

Wood, G. (2018). *Ethereum: A secure decentralised generalised transaction ledger Byzantium version*. Ethereum Project Yellow Paper. Retrieved from https://github.com/ethereum/yellowpaper

Yafimava, D. (2019). Blockchain and logistics: Can the new technology reinvent an outdated system? *OpenLedger*. Retrieved November 10, 2019, from https://openledger.info/insights/blockchain-logistics/

Zhang, R., & Preneel, B. (2017). Publish or perish: A backward-compatible defense against selfish mining in Bitcoin. In H. Handschuh (Ed.), *Topics in cryptology – CT-RSA 2-17: The cryptographers' track at the RSA conference 2017*, San Francisco, CA (pp. 277–292). Cham: Springer. Retrieved from https://www.springerprofessional.de/en/publish-or-perish-a-backward-compatible-defense-against-selfish-/11987450

Chapter 4
Balancing Security: A Moving Target

Artemij Voskobojnikov, Volker Skwarek, Atefeh Mashatan, Shin'Ichiro Matsuo, Chris Rowell, and Tim Weingärtner

4.1 Introduction

4.1.1 Security

In general, security is a non-functional but essential requirement of any IT system. It is also applicable to any IT system based on blockchain technology. However, with blockchain technology, we must differentiate between the usual security of

A. Voskobojnikov (✉)
Department of Electrical and Computer Engineering, University of British Columbia, Vancouver, BC, Canada
e-mail: voskart@ece.ubc.ca

V. Skwarek
Department of Industrial Engineering and Management, Hamburg University of Applied Sciences, Hamburg, Germany
e-mail: volker.skwarek@haw-hamburg.de

A. Mashatan
School of Information Technology Management, Ryerson University, Toronto, ON, Canada
e-mail: amashatan@ryerson.ca

S. Matsuo
Department of Computer Science, Georgetown University, Washington, DC, USA
e-mail: Shinichiro.Matsuo@georgetown.edu

C. Rowell
Sauder School of Business, University of British Columbia, Vancouver, BC, Canada
e-mail: christopher.rowell@sauder.ubc.ca

T. Weingärtner
Lucerne School of Information Technology, Lucerne University of Applied Sciences and Arts, Lucerne, Switzerland
e-mail: tim.weingaertner@hslu.ch

V. L. Lemieux, C. Feng (eds.), *Building Decentralized Trust*,
https://doi.org/10.1007/978-3-030-54414-0_4

blockchain technology as an IT system, and the requirements for securing the "crypto-assets", such as Bitcoin, that blockchains seek to secure.

Definitions of blockchain vary and are still in formation. From the original Bitcoin paper written by Satoshi Nakamoto, the technology was explained as "*an electronic payment system based on cryptographic proof instead of trust, allowing any two willing parties to transact directly with each other without the need for a trusted third party*" (Nakamoto 2008). In this chapter, we define a blockchain as a solution to unauthorized changes to data integrity and to the double-spending problem in a distributed system, using a peer-to-peer distributed timestamp server to generate computational proof of the chronological order of transactions.

The underlying mathematics of Bitcoin is summed up by the definitions above. That is, blockchain technology assures only the chronological order of transactions without a trusted third party. This is the core and mandatory security requirement of a blockchain. On the other hand, other security requirements may be desirable; for example, protection of the privacy of identity regarding each transaction, and the confidentiality of transaction data. The former is partially satisfied by the original Bitcoin implementation, the latter however is not. Thus, these requirements are optional.

In general, security requirements depend on the specification of each IT system. Therefore, when we discuss blockchain security, we need to be careful about the relationship between what we need and what blockchain technology provides. Unfortunately, blockchain technology is not a silver bullet for security issues in our networked society. As described above, the security that this technology offers is incomplete in terms of generally accepted fundamental requirements such as confidentiality, integrity, and availability (CIA) (Stallings and Brown 2018). Thus, we need additional controls for enhanced security.

Blockchain technology implicitly has many (sometimes hidden) assumptions in its operations. Designers of cryptographic protocols do not make general claims such as "all private key (signing keys in the case of blockchains) should be kept secret" because in real-life settings it is very hard to accomplish. Many incidents have led to hundreds of million-dollar losses from cryptocurrency exchanges caused by the difficulty of key management. We need to be aware that cryptography is not the root of trust for *confidentiality*, *integrity,* and *availability*. It is merely the mathematical tool to transform the confidentiality, integrity, and availability problem to the "key management" problem.

Similarly, there are many assumptions in the core blockchain protocol itself. To make the distributed consensus secure and to avoid control by one entity, we need a well-distributed and large enough number of nodes. This number of nodes should also be sustainable. For Proof-of-Work blockchains, poorly distributed hashing power can result in arguably one of the most prominent attacks, the 51% attack. An uneven hashing distribution would allow the group of miners with 51% or more of the overall hashing power to omit new transactions or double-spend coins (see Sect. 4.2.2 for more details).

From the above, when considering the security of blockchains and associated applications, we need to be cautious about what the technology offers, what the assumptions are, and how we can assure requirements that the mathematics of blockchains do not cover. This type of security-by-design thinking should align with the existing Information Security Management System (ISMS: ISO/IEC 27000) framework.

4.1.2 *Trust in an Untrusted Environment*

Blockchain technology is colloquially considered as “trust-free”, or “trustless”, implying the lack of a central governing authority. This notion however can be misleading, as “trustlessness” is often merely a responsibility shift. Transaction partners do not have to trust one another, as long as both share the common belief that the underlying technology will perform as expected. This shifts trust from other places (e.g., identity and access management) to trust in the mathematics of a given blockchain system.

Trust by itself is multi-faceted, and definitions depend on the context and field of study. Two conceptualizations of trust are prevalent in literature, with one being an expectation of a certain behavior in relation to an interaction partner, and the other being seen as willingness to be vulnerable (Beldad et al. 2010). For offline interactions, where people or groups are the interaction partners, the concept of trust might appear obvious. For online interactions however, and blockchain-based systems in particular, the interaction partners might appear intangible and we therefore have to consider how trust is being developed and maintained. Here, social trust between interaction partners is often the result of trust in the technology.

Depending on the type of blockchain, different effects on trust can be expected. Permissioned blockchains offer a certain level of clarity when it comes to interaction partners and their responsibility. Clear governance guidelines and defined roles can alleviate concerns—such as dishonest or malicious actors—associated with public, permissionless blockchains. Here, we have to ask ourselves how to establish trust in an environment that includes actors operating in bad faith, including fraudulent start-ups and exchanges. While transparency can certainly have a positive effect on trust, other antecedents need to be explored that can help in creating the distinction between good and bad actors, which is of utmost importance. Such signals of trustworthiness are emitted by the trustees and create a context in which expectations are being formed by the trustor. Signals are categorized into symbols and symptoms (Riegelsberger et al. 2005) and vary in regard to the degree of reliability they provide. Traditionally, symbols are trust badges, but can also include reputation systems. Both can be easily mimicked by untrustworthy actors, but only if the perceived benefit exceeds the cost of emitting said symbol. Symptoms, however, are generally seen as a by-product of trustworthy actions, and are usually costly to mimic. A large, open-source code base could, for example, emit trustworthiness, as could a large customer base. The latter, however, has to be viewed with caution, as the pseudonymity of blockchain transactions can be leveraged by dishonest actors to artificially alter symptoms indicating growth. As a case in point, unregulated exchanges inflated their trading volumes by up to 95% to signal that the market was stronger than it actually was (Blockchain Transparency Institute 2018).

"Trustlessness" therefore is not a fitting term to gauge the complex trust relationships spanning over social, data/records, and technical layers of a blockchain environment. These relationships are influenced by the stakeholders, their needs, as well as operating contexts, and, moreover, interpretations of whether a system is trustworthy might vary depending on the application area.

4.1.3 Privacy on Blockchains

On a broad level, privacy is considered a basic human right, as recognized in the United Nations Universal Declaration of Human Rights (1948). The definitions are wide-ranging and often include the right to be left alone and the freedom of association. With the rise of big data and emerging technologies such as artificial intelligence and cloud computing, the focus has shifted towards appropriate use and storage of the personally identifiable data of customers. The appropriateness of data usage must be in accordance with laws and policies applicable in the jurisdictions in which a given organization (e.g., a cloud service provider) operates. Naturally, the requirement to abide by laws and policies governing privacy and data protection also holds true for blockchain technologies.

When storing or handling personally identifiable information (PII) in the blockchain context, we have to consider potential implications the underlying technological features might have. In particular, blockchain technology's decentralized nature, as well as the immutability of ledger records, might pose challenges for compliance with regulatory measures such as the General Data Protection Regulation (GDPR). GDPR includes provisions concerning the "right to be forgotten", which refers to the right of an individual to have personal information removed from public access, such as in the case of information available through an internet search (GDPR 2016; Lee 2016). Designing for, and the implications of, privacy and data protection regulations differs depending on the type of blockchain, i.e., permissioned or public, and both types have to be addressed.

For both public and permissioned blockchains, privacy considerations are imperative for system design. Decisions about what data is being stored on-chain can have grave implications for both companies and end users, depending on the application area. Blockchains are immutable by design, and in order to revert a transaction a consensus has to be reached and all participating nodes, whose number can be in the thousands, have to alter the respective local copy of the ledger. For instance, the removal of previously published PII could only be accomplished on public blockchains via achieving a consensus among all nodes, which is costly and might even be infeasible in certain cases. For permissioned blockchains however, where the number of governing nodes is comparatively small, such changes would require less effort. Similarly, updates, e.g., in the case of future regulatory restrictions, could be applied rather seamlessly in the permissioned network, given clear governing guidelines.

Privacy considerations should therefore be factored into the early design stages of the respective blockchain system, as ex-post changes often become increasingly costly and complex with time. More importantly, however, one might argue that transaction reversal, while theoretically feasible, undermines one of the core principles of blockchains, namely immutability, and should only be considered as a last resort.

4.1.4 Security as a Moving Target

Security, while widely labeled as crucial, is merely an afterthought in many cases when it comes to actual system design and implementation. Compared to centralized systems, where the attack surface is limited, blockchains can contain thousands of nodes. This dramatically increases the attack surface and makes blockchains a very attractive target. Attacks on endpoints could further have effects on the whole network, possibly resulting in the entire ledger being compromised. Reactive approaches can therefore only go so far in securing a blockchain system, especially when dealing with an ever-evolving technology; consequently, our attention has to be turned towards a security-by-design paradigm.

It cannot be overemphasized, therefore, that as important as the security-by-design approach might be for conventional systems, it is even more so for the blockchain domain. Preventing bad or vulnerable code is critical when dealing with immutability, and enforcing best practices, such as continuous testing and documentation, can help in achieving these goals. Best practices, however, can only reduce the risk of certain threats; others, such as the threat to conventional cryptography by scalable quantum computers, need to be assessed on a case-by-case basis.

While attacks on the distributed ledgers themselves are arguably more prominent, the overwhelming majority of exploits are caused by complacent end users. With rapidly evolving technologies, end users are constantly facing new challenges when interacting with blockchain-based technologies, and such challenges can result in dangerous errors leading to system failure in the worst case. Solution architects therefore not only need to account for technical vulnerabilities but also for the human factor, which is often considered as the weakest link.

4.2 Security Landscape

4.2.1 Attack Surfaces and Adversarial Goals

Prior to providing an overview of attacks and undertaking threat modelling, we have to first consider attack vectors of blockchains. Generally, an attack vector against an information system is defined as a path or means by which an attacker can gain access to a computer or network server in order to deliver a malicious outcome (ISO 2012, 4.10). An attack surface then is defined as the combination of all attack vectors enabling the adversary in impeding the CIA security principles discussed earlier, which are extended in common security research to CIAAA (*Confidentiality, Integrity, Availability, Authenticity,* and *Accountability).*

It is an information security priority to reduce the attack surface as much as possible in order to counter the adversary. In the context of blockchain technology security, it is therefore imperative that we first examine the attack surface and clearly understand the adversarial model, to be able to better position ourselves against potential security threats (including threats to confidentiality that would negatively affect privacy).

Attack surfaces can be divided into three main categories: network-based, software-based, and user-based. Blockchain technology has vulnerabilities in all of these three categories, as described next.

4.2.1.1 Network-Based Attacks

Blockchains have an inherent peer-to-peer design and therefore are vulnerable to traditional network-based attacks such as distributed denial of service (DDoS) attacks and denial of service (DNS) attacks, e.g., DNS spoofing, where altered DNS records are used to redirect traffic (Pfleeger and Pfleeger 2002). Other blockchain-specific attacks that are carried out via the network surface include the Eclipse node isolation attack (Heilman et al. 2015); Block Withholding, referring to the decreasing of block revenue in a mining pool (Rosenfeld 2011); and Finney Attacks, which are a variation of a double-spending attack (Finney 2011).

4.2.1.2 Software-Based Attacks

The attacks that are made possible due to vulnerabilities introduced by the components of the actual blockchain structure are grouped here. This includes vulnerabilities of the consensus algorithm, as well as the underlying cryptographic primitives used in the implementation of the software. The most well-known attack against blockchain technology so far has been the 51% attack, which is carried out to manipulate the consensus mechanism by controlling more than half of the voting power, e.g., half of the mining power in Proof-of-Work consensus. All permissionless blockchains in operation so far are suffering from their consensus mechanism's weakness against this attack.

Attacks against the underlying cryptographic techniques used in blockchains are also considered a software-based attack. Quantum computing is going to reduce the effective security level of hash functions by a factor of two by means of Grover's search algorithm, which allows searching unsorted databases efficiently (Grover 1996). More importantly, Shor's algorithm (Shor 1994), which addresses the factorization problem, is going to catastrophically break the security of digital signature schemes in current use. Hence the cryptographic techniques and digital signature schemes currently used in blockchain technology must be examined and redesigned to be made quantum resistant. Proposals are currently being examined by the National Institute of Standards and Technology (NIST).

4.2.1.3 User-Based Attacks

Blockchains are very attractive targets for attackers when ordinary human beings, without a lot of security training, are sitting at the endpoints. Many blockchain users do not adhere to proper key management given it is very cumbersome to do so, resulting in, for example, thefts of cryptocurrencies from cryptocurrency exchanges or loss of access to crypto-assets (Voskobojnikov et al. 2020). Cryptojacking is a common attack vector used to exploit the computational power of a target's computer for mining purposes. The open aspect of some platforms that accept smart contracts can allow for malicious code to be introduced and executed. In this case, the immutability of the code on the blockchain can be problematic. Introduced vulnerabilities cannot be fixed as smart contracts are immutable by design. Code audits therefore become of utmost importance prior to deployment.

4.2.2 Technical Weak Points

Having defined the attack surfaces, we can now cover common attacks in more detail. As for all systems carrying and transferring assets [defined as anything that has value to an individual, an organization or a government (ISO 2012)], blockchain technology may be subject to malicious attacks. For a better understanding of attack types and techniques, in the following section we classify and structure known attacks and vulnerabilities. The structure can be made according to different dimensions, using different models: For security considerations, it is rather common to differentiate attack levels by their protocol layer, according to the ISO/OSI-7-layer-model (Zimmermann 1980), or to the more straightforward internet protocol suite (Braden 1989). The internet protocol suite is more abstract than the ISO/OSI-7-layer-model and the latter's seven layers to four (see Fig. 4.1).

However, as blockchains are currently organized as internet applications, although there is potential for much deeper integration it is nearly impossible to assign blockchain functionality over more than the upper protocol layers. Therefore, it is more convenient to use another architectural model for blockchain technology. While many approaches for layered architectures exist (e.g., iFour Technolab Pvt. Ltd. 2019; Javeri 2019; Er-Rajy et al. 2017), some of these architectures are problem-specific, others model workflows instead of layered architectures. Consequently, none of the known approaches is suited as a reference structure for security consideration. Nevertheless, a comparison of these models leads to a consensual number of structural components, as illustrated in Fig. 4.2.

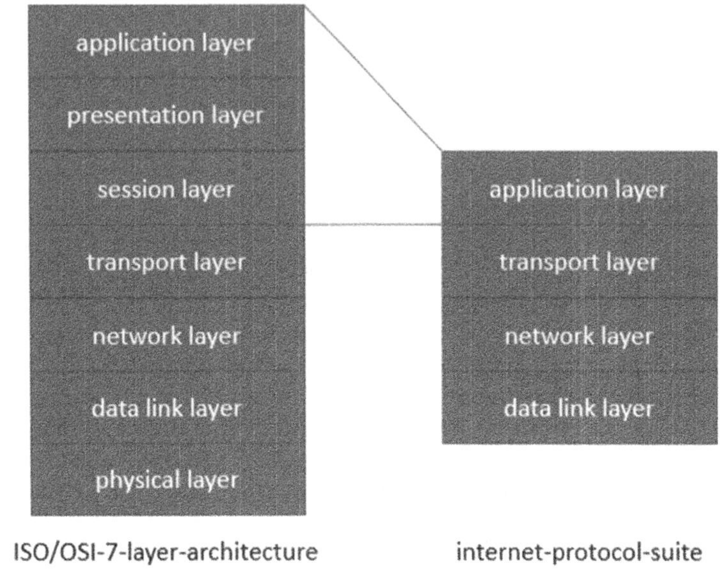

Fig. 4.1 Correlation between the ISO/OSI-7-layer-architecture and the internet-protocol-suite (adapted from Braden 1989)

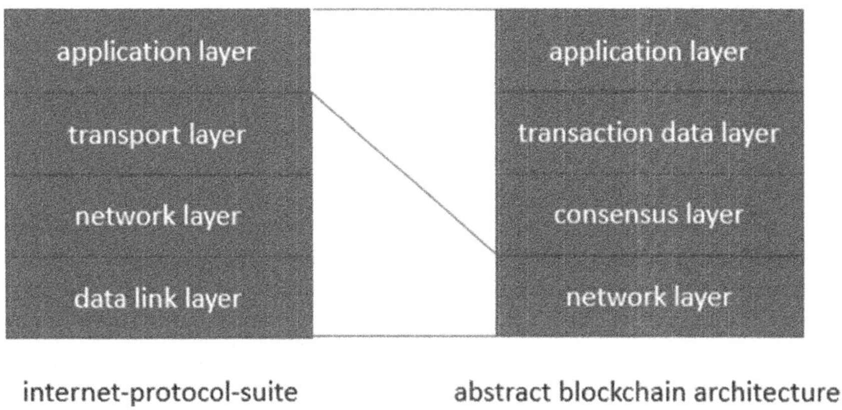

Fig. 4.2 Correlation between the internet-protocol-suite and a potential abstract blockchain architecture (adapted from Fig. 3, Wu and Tran 2018)

The four abstract blockchain architecture layers shown in Fig. 4.2 may be described as follows:

- **Application layer**: applications from in- or outside a blockchain system that are creating or working with transaction data on the blockchain, including blockchain programs such as smart contracts;

- **Transaction data layer**: all kinds of network data as created, transmitted, and received by users of a blockchain system, including code or binary data for blockchain programs such as smart contracts;
- **Consensus layer**: a mechanism for and communication about the correctness of the data–block data layer: all data required for the operation of a blockchain system including user-generated network load data such as addresses (=the value that identifies accounts participating in a transaction), transaction data, as well as system data such as hashes, block numbers or timestamps; and
- **Network layer**: the underlying peer-to-peer-network and associated package data units (PDUs).

This proposed structure offers a framework for the classification of blockchain attacks in the following paragraphs.

Weaknesses can be generally distinguished into those from inside and from outside of the blockchain system. Internal weaknesses, as in all software systems, might be due to software errors. Even in a distributed software system like a blockchain, the source code comes from one single source. Undiscovered errors are distributed to all nodes and therefore are active on all nodes. This might also be used as an attack vector. A number of the recent exploits associated with smart contracts can be attributed to this weakness (Batista and Lemieux 2019).

The more unpredictable technological weaknesses come from outside of the blockchain. If we assume that information added to the blockchain is secure, who guarantees the truth of the origin of this information? Especially if this information comes from an insecure source such as single sensors or manually added data. This can be especially problematic given the immutability of such data once recorded in the blockchain and the assumptions about trust that such recording can imbue.

The most famous attacks affecting the general public occur on the application layer. Security breaches at the interface level, such as breaches of blockchain wallets, are outside of the blockchain but affect the functionality and security of the whole system. Another weakness might come from software running inside the blockchain such as smart contracts. Though smart contracts are not directly linked to the code of the blockchain and should not affect the execution of the blockchain, smart contracts themselves can hold severe errors. A prominent example is the Decentralized Autonomous Organization (DAO) attack from June 2016, which became famous because Ether worth several million US dollars were transferred from a smart contract to another by exploiting a vulnerability in the source code (Atzei et al. 2017). Other examples affect outside-blockchain applications such as trading platforms by attacking their databases and key storage. A very comprehensive work that examines security issues of public blockchains was conducted by Li et al. (2018). This chapter extends their findings and sorts them systematically according to the layered structure as proposed above.

In the following subsections, potential attacks on different layers are listed without a claim of completeness. Ongoing research is needed, therefore, to extend and validate our taxonomy.

4.2.2.1 Attacks on the Application Layer

The most famous are *wallet or exchange hacks*, where attackers gain access to private keys and initiate malicious transactions. Alternatively, they at least manipulate the software behavior so that they can use it on behalf of the user. Additionally, intentionally or accidentally *erroneous smart contracts* may lead to unintended vulnerabilities such as investigated by Atzei et al. (2017).

4.2.2.2 Attacks on the Transaction Layer

Attacks on the transaction layer can be considered as sending corrupted net data. However, at least for data created at the user level, such an attack can be either easy or nearly impossible (=hard) to detect:

- **Simple**: an asset transfer on the blockchain refers either to a direct change (world-state-model such as in Ethereum) or to a relative change (UTXO-model) of the user state. In the case of double-spending or malicious data access, more than only incorrect data is sent; such an attack also requires that user signatures or mining processes be manipulated.
- **Hard**: in the case of user-generated data, this data is a trust anchor in that it is assumed to be accurate (which may be a false assumption) and there is no way for the blockchain to determine the correctness of the data. A singular attack on this level may affect oracles, transferring data from outside to inside the blockchain system. A corrupt oracle may manipulate the data while moving them between systems.

4.2.2.3 Attacks on the Consensus Layer

Forking attacks are most prominent at the consensus layer. 51%-attacks lead to the same consequence as *feather forking* or *punitive forking*: they change a once-met and fixed consensus within the blockchain and directly attack a trust-creating property—its immutability. Also, *selfish mining*, by producing valid blocks without publishing them, can be added to this category. As it is the goal of selfish mining to achieve the longest chain, it is hard to prevent another chain from being created in parallel as long as it is shorter than the secretly selfish mined chain. As soon as the selfish chain is ensured to be (by far) the longest chain, it may also force a fork and invalidate the parallel chain.

4.2.2.4 Attacks on the Block Data Layer

Rare but known attacks on the transaction layer are related to stolen or recovered private keys to create valid address-signature pairs for malicious transactions. In a non-peer-reviewed but often-cited research paper by Mayer (2016), potential vulnerabilities in the ECDSA-key-generation scheme on a theoretical mathematical level were covered. Neither real attacks nor even their attempts have been shown or proven by this paper, but they leave at least a potential of recovering private keys more efficiently than brute force attacks. Most problematic is the immutability-by-concept of the transaction layer. Although the length of the keys is still considered secure, it may be hacked in some near future, and this opens the full history of a blockchain without the chance to recode it to a higher level of security.

Although *double-spending*—creating multiple independent transactions referring to the same base transaction—should be detected by the consensus layer, it is nevertheless considered as an attack because it takes some time before an invalid transaction can be filtered out. Additionally, it is hard to determine which of the multiple transactions should be accepted as authentic and be processed: within the same chain, probably the second transaction would be considered as invalid. However, if both are recorded in parallel chains, the longer chain will win. This attack generally affects multiple layers. However, to be successful, it requires some specialized knowledge about block-building, wallets, and structural overhead information.

Smart contracts may also be subject to attacks on the block data layer in terms of *transaction order attacks*: As a transaction changes the blockchain to a new state, their order of execution on the same smart contract may affect the outcome of the smart contract execution. For example, the ownership of a smart contract may be hijacked by deploying a smart contract to a blockchain. This can occur within the same block-epoch but at an earlier point of transaction handling when an asset-transaction is started from the same address that the smart contract will obtain after its deployment. In this case, the user with the earlier transaction officially owns the address of the smart contract and, therefore, also the smart contract itself.

4.2.2.5 Attacks on the Network Layer

As blockchain systems require a network for their distributed communication, they usually use internet protocol-based systems. Therefore, they are generally vulnerable to all attacks that can damage and exploit internet communication. Most famous attacks based on the routing hierarchy either abuse a monopoly position of a central routing/service instance or reciprocally damage it.

The abuse of a monopoly position is, for example, an *eclipse attack*. Here, an attacker isolates network participants by blocking their internet messages. Principally, this attack should be impossible in a hierarchy-less, distributed network; however, major blockchains such as Bitcoin or Ethereum have been known to suffer from these attacks. As an initialization- and fallback-method for message-routing via peer-nodes fixed neighbor-node tables were used (Bitcoin Core 0.11 (ch 4): P2P network 2018; Chen 2018; Leffew 2019). This was replaced with a regular update of this table and the application of the Kademlia protocol (Maymounkov and Mazières 2002), creating more information about the network neighborhood and distributing the traffic more randomly. Nevertheless, this vulnerability still exists as the list of the initial communication nodes is still required. If the nodes are corrupt, they can divert communications to a corrupt subnetwork, blocking new node traffic.

A counterattack to a single routing node may occur by *flooding*. Independent of the actual technique and protocol layer (data flooding, syn/ack-flooding), this leads to a temporary *denial of service (DoS)* of the participant nodes until routing tables, syn/ack-lists or similar are recovered by time-outs.

4.2.3 Records Weak Points

Blockchains are designed to operate to create trust between social actors (or technical components operated on behalf of social actors, e.g., Internet of Things (IoT) devices) through enabling the creation of trustworthy records of transactions (e.g., ledger records). Underlying weaknesses in the operation of blockchain systems, such as those discussed in the previous section, can compromise the trustworthiness of ledger records. However, as records are different than data, whilst at the same time being comprised of data, in that they often serve to convey important societal rights and entitlements and provide evidence of significant social and business decisions and actions, there are additional specific requirements needed to assure that blockchains are designed to produce trustworthy records. In archival science, records are said to be trustworthy if they are accurate, reliable, and authentic (Lemieux et al. 2019).

Accuracy concerns precision, correctness, truthfulness, and pertinence (Pearce-Moses 2018, s.v. Accuracy). As noted previously, these properties can all be adversely affected if, for example, an inaccurate external data source is used in the creation of a ledger record, such as in the case of a corrupt oracle.

Reliability relates to adherence to formal procedures in the creation of records, completeness of the records in relation to those procedural rules, and the competence of the creator to create the records (Pearce-Moses 2018, s.v. Reliability). A number of aspects of records reliability depend on a determination of how reliably a blockchain system was operating at time of creation, but other aspects of records reliability can only be determined with reference to the legal, administrative and procedural context of the application of the blockchain system to a given use case.

Finally, *authenticity* concerns the ability to determine that the ledger record is what it purports to be (Pearce-Moses 2018, s.v. Authenticity), and requires an unambiguous identity of the record and its creator, and the ability to ascertain that the record has integrity (remains unchanged from its original instantiation). Blockchains excel at integrity, but very often fail to deliver identity of records and their creators. While unique transactions can be used as content addresses for blockchain transactions, they seldom create a bond between the data comprising the transaction and the legal, administrative or procedural purpose of that transaction. When an immutable bond [the "archival bond" (Duranti 1998; Lemieux and Sporny 2017)] is not instantiated in blockchain systems, over time it may become impossible to prove that a given ledger record serves as proof that an ownership right or entitlement was conferred by the record, or for such records to serve as proof of a decision or action. This is because knowledge of the *context* of the ledger record will only be known to and determinable by those who created the records in the first place and will likely recede with the mists of time (and the failings of human memory). Additionally, confirming the identity of records creators is challenging in public, permissionless blockchains that do not require identity to carry out transactions (e.g., they operate pseudonymously), thus making it hard to determine that a ledger record authentically represents the will of a given social actor (e.g., the social actor's will to transfer a certain amount of cryptocurrency, or a cryptoasset, such as ownership of land). Both of these challenges can prevent the realization of accountability in the CIAAA model.

In addition, all of these features needed to instantiate and secure records must be made to persist over time, which requires the application of techniques of digital records preservation. Not only are these techniques not designed with decentralized technologies in mind, they also require frequent migrations to new software that can interfere with the bit-wise integrity checks of blockchain systems. New approaches, such as that being developed by the UK's ARCHANGEL project in collaboration with the UK national archives that uses AI-approaches to determine "allowable" changes in bit-wise integrity of records, could point the way to possible solutions to this conundrum (Collomosse et al. 2018).

4.2.4 Social Weak Points

The blockchain domain is rapidly evolving and is predominantly driven by technological innovation. It is therefore not surprising that both the data/records and technical layers receive far more attention than the social layer in the context of security considerations—leading to users having to adapt to existing software—and less so to software being designed with the users' needs in mind. User-induced errors are prevalent and are often exploited by attackers, more so than the underlying technology itself including the provably secure cryptography.

In traditional online systems the user is exposed to a wide range of threats as discussed above, including but not limited to *phishing*, *malware*, or *man-in-the-middle attacks*. The relevance of these threats becomes evident whenever the confidentiality of credentials is at stake. For instance, in the case of authentication/authorization schemes, such credentials are commonly used for access control, whether the asset in question is an online banking account or a cryptocurrency wallet. Focusing solely on commonalities would however be unjust, as blockchains present unique risks and challenges with which end users are directly or indirectly confronted.

Key features of blockchains, such as immutability and decentralization, are perceived as favorable by many, but can also lead to dangerous errors at the same time. Transactions are irreversible by design, implying that given no centralized authority, the user is fully responsible for their actions. Comparable systems, such as online banking or e-commerce, provide support in case of self-induced errors, and nowadays this is considered as a norm. For blockchains, however, there is no safety net. The user is solely responsible for lost seed phrases or wrongfully-addressed transactions. Errors on the social layer therefore become of utmost importance and the responsibility shift needs to be conveyed clearly to end users. Given the severe impact such errors might have, many companies resort to designing centralized solutions, thus making themselves at least partially responsible in case of user mistakes.

Depending on the user base and application area, finding such a balance between the degree of decentralization and responsibility might become critical. Advanced users might be able to withstand a higher cognitive load when interacting with a system, whereas novice users might surrender when facing even the smallest usability challenges. Other members of society may be incapable of the cognitive effort needed, and regulatory frameworks may be needed to address such situations. Leveraging technological innovation without the users in mind will fail and user experience should no longer be considered a secondary goal, even in technology-driven domains such as blockchain technology-driven innovation.

4.2.5 Failure in Governance: Regulations and Regulatory Goals

In the real world, the use of blockchain technology could be against the social order. For example, many cryptocurrencies were and are currently used for money laundering. This raises the question of how to promote use of blockchain technology that improves the social order rather than undermining it. This is where regulations can prove to be warranted.

Originally, regulations are decided from regulatory goals. According to economic theories, regulatory goals prevent "market failure", which entails preventing crime or enabling consumer protection, and financial stability. These goals are general and, of course, applicable to blockchain-based IT systems. Regulators, however, have faced challenges in responding to the pace of blockchain innovation. The original Satoshi Nakamoto paper was published in 2008, and right after that the reference source code was provided to the public in an open source development style. This caused issues in coordinating regulatory goals and applying regulations to actual implementations of blockchain-based IT systems.

In the history of the development of internet technology, the underlying technology and mathematics come from academic research. The development of internet technology involved a wide range of expertise in order to make the technology suitable for society. Then, companies created actual implementations. After standardization, which arose from multi-stakeholder discussions, the real business started. This sequence of steps created harmonizations among technology and social order (including regulations).

But, in the case of Bitcoin, the actual business started without verification backed by academia and multi-stakeholders. Security of cryptocurrency exchanges is one of the issues caused by such shortcomings. It touches one of the core issues of concern to regulators; that is, consumer protection. Blockchain engineers have not always considered the requirement for consumer protection. Regulators do not have a common language to talk with the open-source engineers. Business entities try to use immature technology to handle hundreds of millions of dollars, and venture capitalists force companies to start their business as early as possible without a truly mature technology. For consumer protection, transparency to the consumer is the crucial aspect; however, it is very difficult for the average consumer, and even many experienced investors and engineers, to critically review many so-called "white paper" documents to determine whether they should cast their money into Initial Coin Offerings (ICO). In general, it is too difficult to judge if specific source code is sufficient from a consumer protection point of view. This is a major missing element in order to rely on the security of blockchain-based systems as a true social foundation.

4.3 The Moving Target: Open Security Challenges of Blockchains

4.3.1 Longevity Requirements for Security of Blockchains

Longevity of security is critical in blockchain security design and implementation to ensure sustainability of the blockchain and its data in the long run. Future threats to the underlying security mechanisms, such as the quantum threat to standardized cryptography and technological obsolescence of blockchain software, and their long-term implications on longevity of blockchains, should be considered and planned for now. The challenge here is to design a system that is going to resist all future attacks and the creative ways adversaries are going to use to try to undermine the security of blockchain systems, as well as be secure against the exigencies of time. This is a near impossible task. Instead, a more practical approach should plan and design for agility so that we can switch between algorithms when a new one is necessary, or migrate seamlessly and without disruption to new software protocols.

We use cryptographic techniques to achieve integrity, authenticity, and confidentiality. Quantum computing may someday defeat critical components of areas of cryptography that are widely used in blockchain implementations. We had the industry-wide SHA1 to SHA2 migration in 2015. Change management aspects of this migration were very costly and time consuming. In the context of blockchain security, the natural question would be to ask: What happens if SHA256 is also deprecated? Another example is digital signatures that are used for integrity and trust in the system. Quantum computers can solve the underlying mathematical problem, i.e., Integer Factorization Problem and Discrete Logarithm Problem (Shor 1994). This breaks our most commonly used digital signature schemes, such as ECDSA and DSA. Although scaleable quantum computers are not yet available, we need to plan for the eventuality that these will be available in the near future.

Any cryptographic migration might entail a fork of a blockchain. Managing forks is not a straightforward task and adds another complexity layer to the longevity requirement of blockchain security.

4.3.2 Regulation, Operation and Security

When considering the security of IT systems, there exist many ISO/IEC and other standards as comprehensive frameworks. For cryptographic technology, ISO/IEC JTC1 SC27/WG2 makes many standards in terms of underlying cryptographic mechanisms. For the verification of cryptographic protocols, ISO/IEC 29128 is the standard to verify and evaluate the level of its security.

To cover the security of hardware/software implementation, ISO/IEC 15408 is the standard to evaluate and certify each product. ISO/IEC 15408 and ISO/IEC 29128 define the levels of certification from a loose to a rigorous level. Each nation has its product certification program which aligns with the ISO/IEC 15408 framework, then certifies each product for use in the nation.

In the records space, ISO/IEC 15489—*Information and Documentation—Records Management* is the predominant standard, while ISO/IEC 30300—*Information and Documentation—Management Systems for Records* provides additional requirements for recordkeeping and ISO/IEC 14721—*Space Data and Information Transfer Systems—Open Archival Information System (OAIS)—Reference Model* provide the basis for long-term preservation of records.

The ISO/IEC 27000 series is well-known as the Information Security Management System (ISMS), in securing operations and lifecycles of IT systems. ISO standards are generally referred to when the government designs any system. This is mandated by the World Trade Organization/Technical Barriers to Trade agreement.

The above is the general and existing regulatory and standard framework in term of security of IT systems. Unfortunately, at the time of writing this book, most of the blockchain implementations and blockchain-based IT systems do not comply with these standards and frameworks. The standards and frameworks are not well-known to young open-source engineers, and the fact that these standards were developed for centralized systems makes them difficult to apply to decentralized systems such as blockchains. Being compatible with these frameworks also requires large budgets that small start-ups cannot cover. However, such standards and frameworks are essential to securing blockchain-based systems and making them transparent to consumers and the government. Standard structures and operations are required for securing blockchain-based IT systems.

4.3.3 Trade-off Between Security and Usability

User experience can be the deciding factor between the success or failure of systems, and balancing security in a way that does not restrict a user's ability to interact with a system is critical. Absolute security and usability are unattainable; the focus should therefore be on a system providing adequate levels of both, given the respective constraints. Past experiences and impressions influence a user's decisions and the more expected and normal a situation appears to be, the more it is trusted. Such *situational normality* is however difficult to attain, particularly for blockchains, where users are constantly confronted with new use cases and terminology. The resulting technology, while innovative, is often hard to use, commonly leading to challenges and errors.

The limited number of user studies in the domain focus on digital currencies and suggest that users appear to be facing hard to overcome usability barriers. It was shown that users of Bitcoin do not necessarily understand the technology, in particular when it comes to privacy implications and the underlying cryptography (Gao et al. 2016). Certainly, one might argue that a user does not necessarily need to understand how the technology works in order to be able to use it, but given the evidence of monetary losses due to self-induced errors (Krombholz et al. 2016; Voskobojnikov et al. 2020) the importance of intuitive software becomes apparent. While these findings are not generalizable to the whole domain, it appears that there is an underlying concern of inadequate mental models, meaning that the user's interpretation of the external reality might lead to dangerous behavior. For example, wallet files might be deleted by unsuspecting users, possibly revoking their access and making the system unusable. Given the wide range of available software, it becomes extremely difficult to define what *usable* actually means in the context of blockchain; thus, we need to look at usability in the general context prior to developing guidelines for the blockchain domain.

Traditionally, usability is defined as the extent to which a user can achieve their goals effectively, but depending on the application such goals can be wide-ranging. Trade-offs between security and usability are therefore contextual and need to be made individually, on a case-by-case basis. Computer security is rarely offered as an option in consumer applications; it is more so a system property that the respective user is not necessarily aware of. Security has to be practically invisible to prevent impediment of workflow efficiency. Notifications, warnings, and options therefore should only be displayed in case of significant risks that the user is exposed to at that point in time, e.g., when making irreversible transactions. The fewer security-critical decisions a user is offered, the fewer potential errors can be made. Here, prioritizing intuitiveness can help in ensuring that existing users do not have to re-learn how to interact with a system and newcomers do not face high entry barriers. Innovative technology can only go so far without usable interfaces and both are equally important in facilitating mass adoption.

Ease of use, or the lack thereof, not only negatively influences existing users but also newcomers who, while eager to learn the technology, are often overwhelmed during the onboarding process. This is particularly interesting for technology adoption where ease of use, among others, has been identified as an influencing factor (Abramova and Böhme 2016). It therefore raises the question of how to design software systems that are perceived as usable by both users and non-users. Here, the subjectively perceived situational normality can be a deciding factor between technology acceptance and rejection on the user's end and naturally, a system designer's goal should be software that is in accordance with past experiences of the respective user. An interesting example highlighting this was an investigation of the unbanked population in Mexico (Larios-Hernández and Ortiz-de-Zarate-Béjar 2019). Besides the lack of trust in institutions, it is argued that participants were accustomed to informal, face-to-face transactions and that blockchain technology could provide an alternative, but only if it would adhere to existing social norms.

It is clearly infeasible to investigate all possible user groups when designing a system; however, taking existing software solutions that already address a given goal as a benchmark can be of help. For instance, cryptocurrency wallets should resemble conventional online banking software and similar analogies can be found for distributed file systems, supply management, and others. For consumer applications technological features should *never* be the main selling point, as the vast majority of users simply would not be able to process such information. User satisfaction hinges on usable interfaces that allow the completion of tasks, and not on the number of buzzwords used in the pitch. Less might therefore be more when it comes to paving the way towards adoption, independent of application area and use case.

4.3.4 Decentralizing Responsibility for Data Security

Blockchain holds significant promise to enhance data security through decentralization. However, as discussed above, there exist inherent trade-offs between security and usability of blockchain technologies for individual users. Given that public blockchain protocols are still in a relatively early stage of emergence, excessive decentralization of responsibility for security could serve to hamper user adoption. Although users could enjoy enhanced security by taking direct control over their personal data, many may actually prefer centralized third parties to hold custodianship of their data and access to this through custodial wallets. By consequence, one of the core benefits of blockchain technologies—enhanced user privacy and security—may be left unrealized, and user adoption in a partially decentralized system could conversely present new security challenges.

In this section, we expand on the security-usability trade-off, and explore how this might shift in the public perception over time. Specifically, we question whether and how users might begin to take security more seriously, and even begin to sacrifice usability and convenience for this. To do so, we situate the emergence of blockchain within a broader trajectory of information governance and cultural awareness, and explore the roles of users, corporations, and hackers in this trajectory. From this, we argue that initially compromising security may enable adoption in the short term; however, creating decentralized solutions may subsequently improve the literacy and practices of these users in the longer term.

To begin, we can simply conceptualize the emergence of blockchain technology along Rogers' (2003) technology adoption curve that depicts the diffusion of innovations. Blockchain-based assets and applications have diffused amongst the "innovators" (the initial 2.5%) and the "early adopters" (the next 13.5%). However, it can take significantly longer for innovations to diffuse amongst the next group, the "early majority", due to this group's different expectations around usability and limited technical literacy. To "cross the chasm" (Moore 2014) between early adopters and the early majority requires innovators to cater to mainstream end users by smoothing the behavioral shifts necessary for adoption.

While this model may be rudimentary, it points to an important qualitative consideration in the diffusion of technologies from tech-savvy users to mainstream adopters: the need to ground innovations in existing understandings. To understand new technologies we tend to lean on comparisons with established products and technologies (Hargadon and Douglas 2001; Navis and Glynn 2010), recruiting metaphor and analogy (Cornelissen and Clarke 2010; Überbacher et al. 2015), and often convey these through narrative and storytelling (Lounsbury and Glynn 2001; Martens et al. 2007; Navis and Glynn 2011; Rosa et al. 1999). Moreover, this is rarely one-directional from innovator to end user, but rather involves dialogue and iteration, in which both groups (along with others such as media organizations) interact and co-create new meanings over time (Navis and Glynn 2010; Rosa et al. 1999).

Applying this to the decentralization of personal data security, we can begin to anticipate challenges by comparing this vision to the existing systems. Over the past two decades we have witnessed the institutionalization of an arrangement in which corporations (such as digital platforms) collect, store, and render their users' personal data. This unique control over user data has become a core part of the business models of many digital platforms (Boudreau and Hagiu 2009; Constantinides et al. 2018). User data and metadata can have "generative" properties for organizations (Tilson et al. 2010; Yoo et al. 2012), meaning that data can be analyzed, aggregated and rendered in ways that help them to improve user experience, create switching costs, and even manipulate user behavior. For instance, in 2014 Facebook purposively filtered user news feeds in a successful attempt to manipulate users' emotional states (Panger 2016). Although this arrangement has been labelled exploitative by some (e.g. Malin and Chandler 2017; Rey 2012; Rogers 2016), individual users arguably benefit by not having to concern themselves with data security. Users can rely upon these organizations when they forget their passwords or accidentally delete their data.

Against this backdrop of an arrangement where organizations assume virtually full responsibility for users' personal data, which remains taken-for-granted and largely unquestioned by users, the complete decentralization of responsibility for data security seems an ominous endeavor. Full decentralization of responsibility is essentially the opposite end of the spectrum to current arrangements. Perhaps then, for blockchain-based assets and applications to break into the mainstream, the early majority may prefer for some dimensions of data security to remain centralized, so that these do not deviate too heavily from existing arrangements.

Such hybrid arrangements appear to be the preferred approach for profit-seeking organizations seeking to promote widespread adoption, for example in the provision of custodial wallets by Facebook (via its Calibra Wallet) in cryptocurrencies and Dapper Labs in online gaming. Especially incumbent organizations may be prone to centralizing aspects of data ownership, custodianship, and/or access as they cling to their existing business models (Barr et al. 1992). However, although centralizing aspects of data custodianship and access may help to forge a path of least resistance to mainstream adoption, such arrangements serve to weaken the security of the system and put individuals at risk. We have witnessed this already in the blockchain space, through the numerous high-profile attacks on cryptocurrency exchanges.

So how could this tension be addressed? Are we destined to continually sacrifice security for usability in ways that blunt one of the cornerstone advantages of distributed ledger technologies? Again, we may be able to shed light on this question by examining the broader trajectory of personal data and its governance. The discourses around how data is collected, stored, and accessed, appear to be shifting in recent years. Whereas in the early- to mid-2000s, the general public was largely unaware that organizations were collecting extensive data about them (let alone that there existed security concerns around this), frequent, high-profile security breaches in recent years have brought data security into the public consciousness. Continued hacks of centralized organizations such as Facebook, Uber, and Equifax, have brought data privacy and security into the forefront of public attention. Those people with a user profile on Ashley Maddison (an online matchmaking service to facilitate extramarital affairs) when it was hacked in July 2017 may understand the importance of data security more than most. In July 2019, Facebook was fined $5 billion US dollars by the Federal Trade Commission over repeated privacy violations, including the Cambridge Analytica scandal in 2016. Together, these high-profile security breaches are bringing data security into focus for the mainstream user.

Bringing these points together, we posit a possible path for the decentralization of responsibility for data security to individual users. In the near-term, centralized organizations might accelerate the adoption process by smoothing the transition for individual users through their hybrid model. This helps users become used to some parts of the new technology whilst masking others. At the same time, however, they establish themselves as "honeypots" for hacks, in the same way that cryptocurrency exchanges have in the past 10 years. Therefore, hacks may actually push people off these custodial wallets and enable them to take full control over their personal data. In short, and counterintuitively, corporations that centralize some parts of data governance may actually be important stepping stones to reach full decentralization.

To conclude, we propose that the history of centralized data ownership and control must be taken into account when projecting the diffusion of blockchain technology and the decentralization of responsibility for data security. However, counterintuitively, the very organizations that have helped to mold existing arrangements may be integral to their replacement, since centralizing data security carries inherent disadvantages and presents them as targets for attack. Over time, we may see a continued trajectory where high-profile security breaches of centralized organizations continue to raise public awareness and literacy around data security in ways that encourage further decentralization.

4.3.5 More Complexity Means Less Security

Complexity—and in this case we mean system complexity—is hard to handle for humans. We experience this every day and it is not only true for the blockchain environment. Let us take a look at several examples where we experience complexity in blockchains and let us use the above-defined layers, or dimensions, to tie the discussion together.

4.3.5.1 Social Layer/Dimension

For the vast majority of the population blockchain technology is a closed book. Even for some people dealing with cryptocurrencies like Bitcoin the underlying technology is not comprehensible.

Wallets are a good example of this. They store the private keys which are needed for accessing blockchain addresses. This is required because the handling of a 64-character key is too complex for humans to remember. Therefore, we use wallets or QR codes to reduce this complexity, but this approach introduces several security issues. What if the wallet application sends the private key to some third person? What if the QR code does not represent the intended address? What if another application pretends to be the wallet app and steals the password of the proper wallet?

Another example is trust relations. Blockchain is said to allow trust even if the counterparty is not known. This is because one can trust the immutability of transactions, the identity behind an address, and the transparency of entries to name a few. But how do we verify this? Do we vet the number of miners and understand their relationship enough to exclude a 51% attack?

And finally, in a legal sense, the complexity of our legal system in combination with a complex technology like blockchain has reached such a high degree that legal compliance for many use cases cannot be guaranteed. Even lawyers are often overwhelmed and have to wait for court decisions to be on the safe side of things.

4.3.5.2 Data Records Layer/Dimension

If we take a look at the complexity-security relation in the data records layer, or dimension, we can observe that trustworthiness of data is a complex issue, too. A good illustration of this issue can be found in the context of IoT devices.

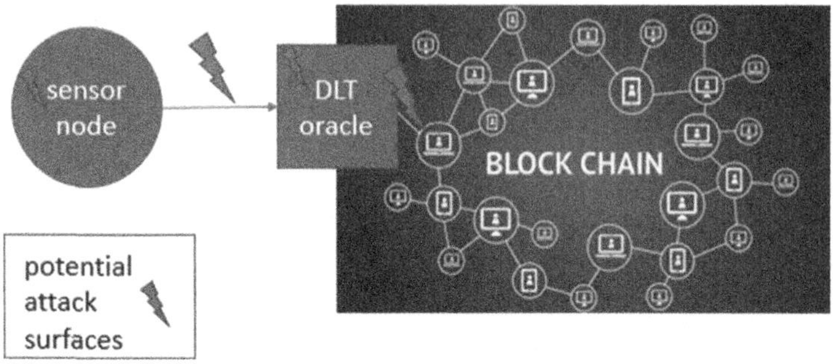

Fig. 4.3 Attack surfaces of sensor connections and DLT-oracles for blockchains

IoT suffers from multiple different definitions. In a very loose and wide definition provided by Farooq et al. (2019), which is itself derived from Singh et al. (2014), the IoT, "provides internet-based services that involve human-to-thing, thing-to-thing, and thing-to-things communications." Therefore, things are involved which are usually considered as sensors or sensor networks. Other terms such as smart fog or smart dust in contrast to high-performance cloud-systems are used (Skwarek et al. 2016).

A basic property of such IoT-devices in terms of sensors, sensor systems, or sensor networks is their simplicity, which is intended to achieve:

- low size;
- low power-consumption;
- low cost; and
- long operating times.

These main design parameters are mostly achieved with low-performance micro-controllers. At the same time, such devices are used as real-world-data sources (=sensors) for trusted systems such as blockchains. This arrangement creates a potential for data trustworthiness issues.

In common setups, the sensors—or IoT-devices—are connected as singular devices via access-points (=gateways) to the blockchain as a data-source (=DLT-oracle) (see Fig. 4.3), sending their data via an (un)trusted software into the blockchain. Given the fact that the sensor data itself must be considered as correct, because this is the defined trust-anchor and will not be questioned (unwisely), the communication channel of the potentially wireless sensor and the operation of the DLT-oracle software might be subject to an attack.

The scenario gets even more complicated when a multi-sensor-network is attached to the DLT-oracle, e.g., for monitoring goods during transport. Many sensors may be distributed among the load sending all their "trustful" data to a wireless gateway, also working as a DLT-oracle server into a blockchain. In this scenario a blockchain is required as the data is required for later inspection (e.g., as evidence of how the shipment was handled). To assure trusted communication, the communication channel is usually encrypted.

The required long battery operating times of the sensors are usually achieved by low power consumption—a combination of low-performing processor capabilities and sleep intervals. Especially the sleep intervals are security-critical regarding the "A" (i.e., availability) of the CIAAA principle. During this time, an attacker is able to execute multiple attack-schemes:

- **Man-in-the-middle-replay**: The attacker can replay earlier recorded data on behalf of a suspended sensor, even if the channel is encoded, as the encryption scheme or -key may not have changed.
- **Sybil-attack**: The attacker could become a silent listener to the decoded channel and generate its own data under the identity of the suspended sensor. In real idle-phases, the suspended sensor may not even realize that its identity has been captured and abused by an attacker.

Many attacks can be listed and considered as sensor nodes and networks are not capable of complex operation modes due to their simplistic design principles. To illustrate in terms of the CIA-triad:

- **Confidentiality**: Complex channel encryption requires too much computational effort and energy at the expense of the operating time, therefore most channels in wireless sensor communication are not highly secured.
- **Integrity**: An integrity check can principally be hash-or checksum-based. But as the checksum algorithm has to be known in order for the check to be performed, the value can easily be generated by an attacker. Consequently, using single sensor values, the integrity has simply to be taken as a trust anchor.
- **Availability**: As already discussed, unavailability is a part of the design-principles of an IoT-network, which opens the door to various types of attacks.

These security weaknesses are not insurmountable, however, as discussed in detail in Box 4.1.

Box 4.1 Securing IoT Sensor Communications in the Context of Blockchains Using the Sensorchain Concept

A higher level of security can be achieved by the application of methods of Byzantine fault tolerance, such as those used in many distributed ledgers. This enables the sensors to become reliable and a trusted part of a network by repeating longer time series of messages and performing redundant checks, whether such messages have been received on redundant communication paths or not. As an example see Fig. 4.4. Such communication principles are known from systems such as Hashgraph (Baird 2016), Iota (Popov 2016) or sensorchain (Skwarek 2017) and evolve from research into practical applications. Short snipplets or time series of past sent and received communication are repeated in a new message to show other participants that the IoT-node knows about some history of communication.

(continued)

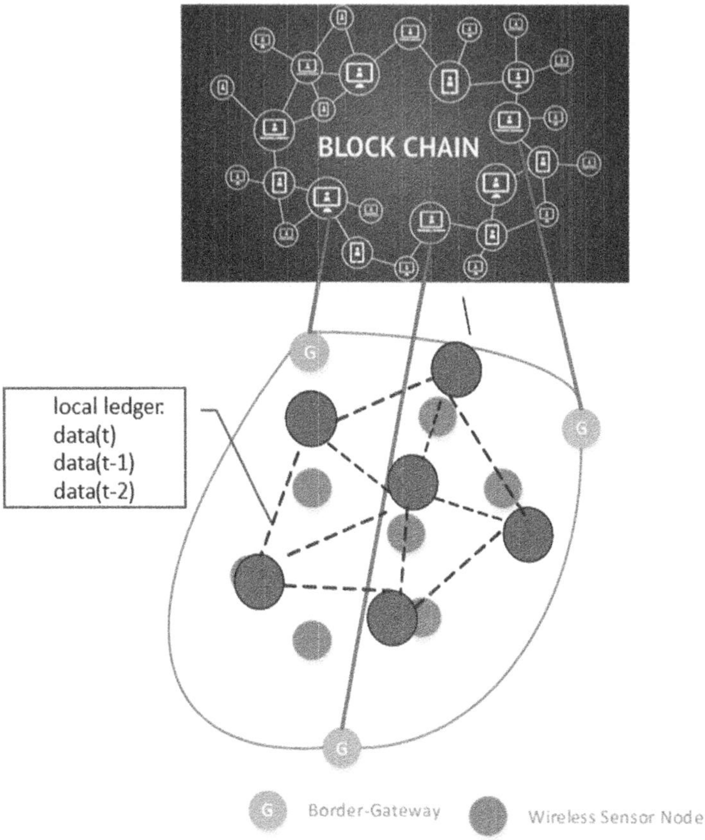

Fig. 4.4 The sensorchain concept

Box 4.1 (continued)

Other participants receiving these BFT-messages are now able to decide about the correctness of the messages in terms of integrity. Moreover, missing availability can be bridged by this method. Although idle-intervals may still lead to data gaps in the time series of BFT-messages, similarities to earlier communication epochs can be detected by other nodes allowing for a means of determining if missing data due to unavailability of a node are plausible and trustworthy.

If the identity of a node is not only generated by some static ID-number, but is generated according to the location of a node such as described by Bornholdt et al. (2019), the identity can also be verified independently by other nodes.

(continued)

Box 4.1 (continued)
Therefore the man-in-the-middle or sybil-attacks can be detected and countered by the network or by the gateway.

Consequently decentralized communication of IoT-systems can be secured by methods of DLT-like BFT protocols.

The anonymity of a blockchain is achieved by avoiding usernames and instead using addresses. The cryptic and often randomly generated address, in combination with using addresses only once, gives us the apparent safety of anonymity. But this pseudonymity only lasts as long as no one can make the link between an address and a user. Once this link is established, former transactions can be viewed due to the immutability of the blockchain. In this case the complexity of the address makes us believe in its security. The desire for data anonymity, moreover, may be in direct conflict with the need we have in some cases (e.g., transfer of property rights) to establish the identity (legal or at least social) of a transacting party to establish the authenticity of records.

One argument for using a blockchain is transparency. In permissionless blockchains everybody can have a look at the data. Therefore, it is said to be transparent. But have you ever had a look at this data? Did you understand the semantics? The maximum degree of transparency we usually check are some webpages showing the transactions and even this is not comprehensible to ordinary citizens. Can we handle this kind of transparency or is it already too complex?

4.3.5.3 Technological Layer/Dimension

On the technological layer or dimension, the complexity of blockchain platforms rises with the use of smart contracts. Those self-executing programs allow control over data and assets. The more complex they become, the more likely an error might be included. Since smart contracts cannot be altered once deployed to the blockchain any error might result in the loss of assets.

Finally, the blockchain platforms themselves are highly complex systems which are understood only by a few. Errors in the code of such platforms cannot be fixed like in ordinary computer applications since they are distributed over many nodes. Each node has to agree on an update and perform this update in the same time period to ensure the functionality of the blockchain. Disagreements over changes result in a fork of the blockchain. In this case the security and longevity of the data (and associated records) cannot be guaranteed anymore.

How should we deal with this dilemma? Can blockchains as a complex system be saved at all? The answer may be in nature. As an adaptable highly complex multi-agent system, nature deals with complexity in an excellent manner. Techniques applied are:

- Distribution and redundancy, which we already use in blockchain technology;
- Adaptation to change, which some blockchain platforms like Tezos are attempting to implement;
- Limited life expectancy;
- Resilience or the ability not to fail completely in the event of disturbances or partial failures, but to maintain essential services; and/or
- No stasis, but evolution and survival of the fittest.

Will we find a way to make our blockchain systems robust and secure even though they may become more and more complex? And can nature be a role model for this?

4.4 The Constant

As discussed throughout the previous sections of this chapter, security in the context of blockchain-based and distributed ledger technologies cannot be generalized. However, despite the wide range of application areas and influencing factors, two themes can be observed that can serve as guidance for security in the context of this rapidly evolving technology. Here, *agile security* becomes an integral part of the technology, particularly when considering how costly changes are. Further, the role of the alleged *weakest link*, i.e., the end users, and their influence on a blockchain ecosystem. Certainly, the two themes are not unique to the blockchain space, however, both become increasingly important due to the inherent properties of the technology, such as immutability and decentralization.

4.4.1 Designing for the Future

Only few could have guessed the rapid development of the blockchain domain since Bitcoin's inception in 2009. With thousands of cryptocurrencies and tokens, various sectors making use of the technology and millions of users worldwide, there appears to be great potential in this technology. However, making predictions of how the future might unfold and how the domain might look in 10 years' time is not the goal of this chapter. Here, the focus is on security considerations that can help in designing solutions flexible enough to meet the requirements of the future.

Traditionally, a security goal refers to an asset and defines the security objective, i.e., what attribute of the asset is at stake and needs to be protected. While we cannot predict what assets and stakeholders might emerge over time, the definition of what confidentiality, integrity, and availability mean will likely remain relatively stable. The means with which these attributes (i.e., the CIA-triad) can be protected will change of course, with quantum cryptography being one challenge in the near future. Rather than tackling security challenges as they come, we argue for pre-emptively creating software that is expandable enough to mitigate issues as they arise in the first place. For public blockchains, such updates come with great cost due to the large number of participating nodes. In this case, updates can only take place once a consensus is reached, and past disagreements resulting in hard forks have shown how difficult this might be. Changes can have grave implications and have therefore to be considered carefully. While disagreements cannot be prevented, update protocols and transparent decision making can help in addressing some of the concerns that have arisen in the case of the biggest hard forks in the past, namely Ethereum and Bitcoin.

For permissioned blockchains however, the argument for expandability is easier to make due to the limited number of participating nodes and their willingness to cooperate. Still, clear governing guidelines are needed to reach consensus and must be created from the very beginning to avoid disagreements later on.

Overall, it appears that in a rapidly evolving domain there is simply no place for stagnant computer security. Risks are changing and so are security requirements. *Agile security* should therefore be a core element of solution design and not merely an afterthought, as is often the case.

4.4.2 The Weakest Link

The human factor is often labeled as the weakest cybersecurity link in both academia and industry but is this assessment truly fair? We argue that the end users are doing their best to adapt to rapidly evolving technology and might simply fall short while doing so. Lowering the cognitive load that users endure during interaction should therefore be the primary goal of solution designers and architects.

The first users of blockchain technology were the select few on Bitcoin forums in 2009. Since then, the user base has grown but has the technology and the user experience also changed? While hundreds of new wallets were developed, the resemblance to the original Satoshi client is clearly there. Users still have to deal with public key cryptography, key management, and confusing terms in interfaces that have already existed more than 10 years ago. While cypherpunks in 2009 were more or less comfortable with this software, it is doubtful that the common computer user nowadays will be as well. Several studies suggest that public key cryptography is hard to use (Whitten and Tygar 1999; Sheng et al. 2006; Ruoti et al. 2015) and this is not surprising. Back in 1999, the average user was struggling with public key cryptography and evidently this is still the case, with the user still being unable to use Bitcoin wallets (Eskandari et al. 2015). Cryptography is hard to grasp and expecting end users to adjust to the technology that is being thrown at them is unjust. While several improvements have been made to enhance user experience, users continue to struggle. High barriers to entry and switching costs are a hindrance to adoption and the technological advancements of blockchains will not matter as long as such barriers exist. If there is no user base, what value is there in a groundbreaking technology? A product becomes successful through its users and user experience.

Great user experience has always been one of the deciding factors in technology adoption, with leading examples from Apple or Facebook. It is hard to disrupt existing technologies when the user experience is lacking. The average user will not choose a platform because of its technological features: a platform will be chosen if it addresses a need without putting exceptional amounts of cognitive load on the respective end user. A technology has to be both *usable* and *useful* to be adopted in the long run and both of these attributes clearly rely on the perceptions of users.

Innovations are adopted over time and while early adopters might already be on board, the early majority is not there yet. Designers should take existing solutions as a benchmark when thinking about interfaces for blockchain technology. If possible, the users should not even be aware of the underlying technology. Usable security research in the space is in its infancy and deserves more attention as technological innovation alone can only partially pave the way towards mass adoption.

References

Abramova, S., & Böhme, R. (2016). Perceived benefit and risk as multidimensional determinants of Bitcoin use: A quantitative exploratory study. In *Proceedings of the 37th International Conference on Information Systems (ICIS 2016)* (pp. 233–252). Atlanta, GA: Association for Information Systems. https://aisel.aisnet.org/icis2016/Crowdsourcing/Presentations/19/

Atzei, N., Bartoletti, M., & Cimoli, T. (2017). A survey of attacks on Ethereum smart contracts (SoK). In M. Maffei & M. Ryan (Eds.), *POST 2017: Principles of security and trust* (Lecture notes in computer science) (Vol. 10204, pp. 164–186). Berlin: Springer. https://doi.org/10.1007/978-3-662-54455-6.

Baird, L. (2016). The Swirlds hashgraph consensus algorithm: Fair, fast, byzantine fault tolerance: SWIRLDS-TR-2016-01. Retrieved from https://www.swirlds.com/downloads/SWIRLDS-TR-2016-01.pdf

Barr, P. S., Stimpert, J. L., & Huff, A. S. (1992). Cognitive change, strategic action, and organizational renewal. *Strategic Management Journal, 13*(S1), 15–36. https://doi.org/10.1002/smj.4250131004.

Batista, D., & Lemieux, V. (2019). Bounded and shielded: Assessing security aspects and trustworthiness of smart contracts. In *Proceedings of the Annual Conference of the Canadian Association for Information Science (CAIS), University of Alberta Libraries, AB, June 4, 2019*. Retrieved from https://journals.library.ualberta.ca/ojs.cais-acsi.ca/index.php/cais-asci/

Beldad, A., de Jong, M., & Steehouder, M. (2010). How shall I trust the faceless and the intangible? A literature review on the antecedents of online trust. *Computers in Human Behavior, 26*(5), 857–869. https://doi.org/10.1016/j.chb.2010.03.013.

Bitcoin Core 0.11 (ch 4): P2P network. (2018). In *Bitcoin Wiki*. Retrieved August 27, 2019, from https://en.bitcoin.it/wiki/Bitcoin_Core_0.11_(ch_4):_P2P_Network#Peer_discovery

Blockchain Transparency Institute. (2018). *December 2018: Exchange volumes report*. Originally retrieved February 19, 2019, from https://www.blockchaintransparency.org/december-2018-rankings (now renamed Market Surveillance Report – December 2018 and available at https://www.bti.live/reports-december2018/)

Bornholdt, L., Reher, J. & Skwarek, V. (2019). Proof-of-location: A method for securing sensor-data-communication in a Byzantine fault tolerant way. In *Mobile communication – Technologies and applications; 24. ITG-Symposium* (pp. 1–6). IEEE. Retrieved from https://ieeexplore.ieee.org/document/8731780

Boudreau, K. J., & Hagiu, A. (2009). Platforms rules: Multi-sided platforms as regulators. In A. Gawer (Ed.), *Platforms, markets and innovation* (pp. 163–191). Cheltenham: Edward Elgar

Braden, R. (Ed.). (1989). *Requirement for internet hosts – Communication layers*. Internet Engineering Task Force: Network Working Group: RFC 1122. Retrieved August 23, 2019, from https://tools.ietf.org/pdf/rfc1122.pdf

Chen, L. (2018, November 15). Peer discovery in Harmony network. *Medium*. Retrieved August 27, 2019, from https://medium.com/harmony-one/peer-discovery-in-harmony-network-6a07f9401c61

Collomosse, J., Bui, T., Brown, A., Sheridan, J., Green, A., Bell, M., Fawcett, J., & Thereaux, O. (2018). ARCHANGEL: Trusted archives of digital public documents. In *DocEng '18: Proceedings of the ACM Symposium on Document Engineering 2018* (Article 31, pp. 1–4). New York: Association for Computing Machinery. https://doi.org/10.1145/3209280.3229120.

Constantinides, P., Henfridsson, O., & Parker, G. G. (2018). Platforms and infrastructures in the digital age. *Information Systems Research, 29*(2), 381–400. https://doi.org/10.1287/isre.2018.0794.

Cornelissen, J. P., & Clarke, J. S. (2010). Imagining and rationalizing opportunities: Inductive reasoning and the creation and justification of new ventures. *The Academy of Management Review, 35*(4), 539–557. https://doi.org/10.5465/amr.35.4.zok539.

Duranti, L. (1998). *Diplomatics: New uses for an old science*. Lanham, MD: Scarecrow Press

Er-Rajy, L., El Kiram, M.A., El Ghazouani, M., & Achbarou, O. (2017). Blockchain: Bitcoin wallet cryptography security, challenges and countermeasures. *Journal of Internet Banking and Commerce, 22*(3), 294. Retrieved August 24, 2019, from http://www.icommercecentral.com/peer-reviewed/blockchain-bitcoin-wallet-cryptography-security-challenges-and-countermeasures-86561.html

Eskandari, S., Barrera, D., Stobert, E., & Clark, J. (2015). *A first look at the usability of Bitcoin key management*, presented at USEC '15, San Diego, CA, February 8, 2015. https://arxiv.org/abs/1802.04351

Farooq, U., Ul Hasan, N., Baig, I., & Shelzad, N. (2019). Efficient adaptive framework for securing the Internet of Things devices. *EURASIP Journal on Wireless Communications and Networking, 2019*, 210. https://doi.org/10.1186/s13638-019-1531-0.

Finney, H. (2011). *The Finney attack (the Bitcoin Talk forum), 2011*. Retrieved April 7, 2020, from https://bitcointalk.org/index.php?topic=3441.msg48384

Gao, X., Clark, G.D., & Lindqvist, J. (2016). Of two minds, multiple addresses, and one ledger: Characterizing opinions, knowledge, and perceptions of Bitcoin across users and non-users. In: *CHI'16: Proceedings of the 2016 CHI Conference on Human Factors in Computing Systems* (pp. 1656–1668). New York: Association for Computing Machinery (ACM). http://doi.acm.org/10.1145/2858036.2858049.

General Data Protection Regulation 2016/679 Article 17: Right to erasure ('right to be forgotten') (EU). Retrieved November 23, 2019, from https://gdpr-info.eu/art-17-gdpr/

Grover, L. K. (1996). A fast quantum mechanical algorithm for database search. In *STOC '96: Proceedings of the Twenty-Eighth Annual ACM Symposium on Theory of Computing* (pp. 212–219). New York: Association for Computing Machinery (ACM). https://doi.org/10.1145/237814.237866.

Hargadon, A. B., & Douglas, Y. (2001). When innovations meet institutions: Edison and the design of the electric light. *Administrative Science Quarterly, 46*(3), 476–501. https://www.jstor.org/stable/3094872/

Heilman E., Kendler A., Zohar A., & Goldberg S. (2015). Eclipse attacks on Bitcoin's peer-to-peer network. In J. Jung (Ed.), *Proceedings of the 24th USENIX Conference on Security Symposium (SEC'15)* (pp. 129–144). Berkeley, CA: USENIX Association. https://dl.acm.org/doi/10.5555/2831143.2831152.

iFour Technolab Pvt. Ltd. (2019, April 11). *Blockchain and architecture* [Blog post]. Retrieved August 24, 2019, from https://www.ifourtechnolab.com/blog/blockchain-history-and-evolution

International Organization for Standardization (ISO). (2012). *Information technology—security techniques—guidelines for cybersecurity* (ISO/IEC 27032:2012)

Javeri, P. (2019). Blockchain architecture. *Medium*. Retrieved August 28, 2019, from https://medium.com/@prashunjaveri/blockchain-architecture-3f9f1c6dac5e

Krombholz, K., Judmayer, A., Gusenbauer, M., & Weippl, E. (2016). The other side of the coin: User experiences with Bitcoin security and privacy. In J. Grossklags & B. Preneel (Eds.), *Financial cryptography and data security* (pp. 555–580). Berlin: Springer. https://doi.org/10.1007/978-3-662-54970-4_33.

Larios-Hernández, G. J., & Ortiz-de-Zarate-Béjar, A. (2019). Blockchain entrepreneurship and the struggle for trust among the unbanked. In H. Treilblmaier & R. Beck (Eds.), *Business transformation through blockchain: Vol. II* (pp. 259–283). Cham: Springer International. https://doi.org/10.1007/978-3-319-99058-3_10.

Lee, J. (2016). What the right to be forgotten means to companies: Threat or opportunity? *Procedia Computer Science, 91*, 542–546. https://doi.org/10.1016/j.procs.2016.07.138.

Leffew, K. (2019). A brief overview of Kademlia, and its use in various decentralized platforms. *Medium*. Retrieved August 27, 2019, from https://medium.com/coinmonks/a-brief-overview-of-kademlia-and-its-use-in-various-decentralized-platforms-da08a7f72b8f

Lemieux, V. L., & Sporny, M. (2017, April). Preserving the archival bond in distributed ledgers: A data model and syntax. In *Proceedings of the 26th International Conference on World Wide Web Companion* (pp. 1437–1443). https://doi.org/10.1145/3041021.3053896.

Lemieux, V. L., Hofman, D., Batista, D., & Joo, A. (2019). *Blockchain technology and recordkeeping* (Report prepared for the ARMA International Education Foundation). Retrieved from http://armaedfoundation.org/wp-content/uploads/2019/06/AIEF-Research-Paper-Blockchain-Technology-Recordkeeping.pdf

Li, X., Jiang, P., Chen, T., Luo, X., & Wen, Q. (2018). A survey on the security of blockchain systems (arXiv:1802.06993). *arXiv.org*. Retrieved August 24, 2019, from http://arxiv.org/abs/1802.06993

Lounsbury, M., & Glynn, M. A. (2001). Cultural entrepreneurship: Stories, legitimacy, and the acquisition of resources. *Strategic Management Journal, 22*(6–7), 545–564. https://onlinelibrary.wiley.com/doi/10.1002/smj.188

Malin, B. J., & Chandler, C. (2017). Free to work anxiously: Splintering precarity among drivers for Uber and Lyft. *Communication, Culture & Critique, 10*(2), 382–400. https://onlinelibrary.wiley.com/doi/abs/10.1111/cccr.12157

Martens, M. L., Jennings, J. E., & Jennings, P. D. (2007). Do the stories they tell get them the legitimacy they need? The role of entrepreneurial narratives in resource acquisition. *The Academy of Management Journal, 50*(5), 1107–1132. Retrieved from https://www.jstor.org/stable/20159915

Mayer, H. (2016). ECDSA security in Bitcoin and Ethereum: A research survey. *CoinFabrik*. Retrieved from https://blog.coinfabrik.com/ecdsa-security-in-bitcoin-and-ethereum-a-research-survey/

Maymounkov, P., & Mazières, D. (2002). Kademlia: A peer-to-peer information system based on the XOR metric. In P. Druschel, F. Kaashoek, & A. Rowstron (Eds.), *Peer-to-peer systems: First international workshop, IPTPS 2002, Cambridge, MA, USA, March 7–8, 2002. Revised papers. Lecture notes in computer science* (Vol. 2429). Berlin: Springer. https://doi.org/10.1007/3-540-45748-8_5.

Moore, G. A. (2014). *Crossing the chasm: Marketing and selling disruptive products to mainstream customers* (3rd ed.). New York: HarperCollins

Nakamoto, S. (2008). *Bitcoin: A peer-to-peer electronic cash system*. Satoshi Nakamoto Institute. Retrieved from https://nakamotoinstitute.org/bitcoin/

Navis, C., & Glynn, M. A. (2010). How new market categories emerge: Temporal dynamics of legitimacy, identity, and entrepreneurship in satellite radio, 1990–2005. *Administrative Science Quarterly, 55*(3), 439–471. https://doi.org/10.2189/asqu.2010.55.3.439.

Navis, C., & Glynn, M. A. (2011). Legitimate distinctiveness and the entrepreneurial identity: Influence on investor judgements of new venture plausibility. *The Academy of Management Review, 36*(3), 479–499. Retrieved from https://www.jstor.org/stable/41319182

Panger, G. (2016). Reassessing the Facebook experiment: Critical thinking about the validity of Big Data research. *Information, Communication & Society, 19*(8), 1108–1126. https://doi.org/10.1080/1369118X.2015.1093525.

Pearce-Moses, R. (ed.) (2018). *InterPARES trust terminology*. InterPARES Trust. Retrieved from https://interparestrust.org/terminology/

Pfleeger, C. P. & Pfleeger, S. L. (2002). *Security in computing* (3rd ed.). Prentice Hall Professional Technical Reference

Popov, S. (2016). The tangle whitepaper. *IOTA.org*. Originally retrieved May 25, 2017, from https://www.iotatoken.com/IOTA_Whitepaper.pdf. Now available at http://www.tangleblog.com/wp-content/uploads/2016/11/IOTA_Whitepaper.pdf

Rey, P. J. (2012). Alienation, exploitation, and social media. *American Behavioral Scientist, 56*(4), 399–420. https://doi.org/10.1177/0002764211429367.

Riegelsberger, J., Sasse, M. A., & McCarthy, J. D. (2005). The mechanics of trust: A framework for research and design. *International Journal of Human-Computer Studies, 62*(3), 381–422. https://doi.org/10.1016/j.ijhcs.2005.01.001.

Rogers, E. (2003). *The diffusion of innovations* (5th ed.). New York: The Free Press

Rogers, B. (2016). The social costs of Uber. *University of Chicago Law Review Online, 82*(1), 85–102. Retrieved from https://chicagounbound.uchicago.edu/uclrev_online/vol82/iss1/6

Rosa, J. A., Porac, J. F., Runser-Spanjol, J., & Saxon, M. S. (1999). Sociocognitive dynamics in a product market. *Journal of Marketing, 63*, 64–77. Retrieved from https://www.jstor.org/stable/1252102

Rosenfeld, M. (2011). Analysis of Bitcoin pooled mining reward systems (arXiv:1112.4980). *arXiv.org*. Retrieved from https://arxiv.org/abs/1112.4980

Ruoti, S., Andersen, J., Zappala, D., & Seamons, K. (2015). Why Johnny still, still can't encrypt: Evaluating the usability of a modern PGP client (arXiv:1510.08555). *arXiv.org*. Retrieved from https://arxiv.org/abs/1510.08555

Sheng, S., Broderick, L., Koranda, C. A., & Hyland, J. J. (2006). *Why Johnny still can't encrypt: Evaluating the usability of email encryption software*. Poster session presented at the meeting of SOUPS 2006: Symposium on Usable Privacy and Security, Pittsburgh, PA, July 12–14, 2006. Abstract. Retrieved from http://www.chariotsfire.com/pub/sheng-poster_abstract.pdf

Shor, P. W. (1994). Algorithms for quantum computation: Discrete logarithms and factoring. In *Proceedings 35th Annual Symposium on Foundations of Computer Science* (pp. 124–134). IEEE. Retrieved from https://ieeexplore.ieee.org/document/365700

Singh, D., Tripathi, G., & Jara, A. J. (2014). A survey of Internet-of-Things: Future vision, architecture, challenges and services. In *2014 IEEE World Forum on Internet of Things (WF-IoT)* (pp. 287–292). Retrieved from https://ieeexplore.ieee.org/document/6803174

Skwarek, V. (2017). Blockchains as security-enabler for industrial IoT applications. *Asia Pacific Journal of Innovation and Entrepreneurship 11*(3), 301–311. https://doi.org/10.1108/APJIE-12-2017-035. Retrieved January 6, 2019, from http://www.emeraldinsight.com/doi/10.1108/APJIE-12-2017-035

Skwarek, V., Kistler, T., Rawer, M., & Schauer, S. (2016). IoT und sensornetzwerke: entwurf und programmierung von niedrigstenergiesystemen anhand einer metaarchitektur [IoT and sensor networks: Design and programming of lowest energy systems based on a meta-architecture]. In H. C. Mayr & M. Pinzger (Eds.), *Lecture Notes in Informatics (LNI), Proceedings – Series of the Gesellschaft fur Informatik (GI)P-259 – INFORMATIK 2016*, (pp. 1917–1925)

Stallings, W., & Brown, L. (2018). *Computer security: Principles and practice* (4th ed.). New York: Pearson

Tilson, D., Lyytinen, K., & Sørensen, C. (2010). Digital infrastructures: The missing IS research agenda. *Information Systems Research, 21*(4), 748–759. Retrieved from https://www.uio.no/studier/emner/matnat/ifi/INF5210/h14/pensumliste/articles/tilson-et-al-2010.pdf

Überbacher, F., Jacobs, C. D., & Cornelissen, J. P. (2015). How entrepreneurs become skilled cultural operators. *Organization Studies, 36*(7), 925–951. https://doi.org/10.1177/0170840615575190.

Voskobojnikov, A., Obada-Obieh, B., Huang, Y., & Beznosov, K. (2020, February). Surviving the cryptojungle: Perception and management of risk among North American cryptocurrency (non) users. In J. Bonneau J. & N. Heninger (Eds.) Financial cryptography and data security. FC 2020. Lecture notes in computer science (Vol 12059, pp. 595-614). Cham: Springer. https://doi.org/10.1007/978-3-030-51280-4_32.

Whitten, A., & Tygar, J. D. (1999). Why Johnny can't encrypt: A usability evaluation of PGP 5.0. In: *SSYM'99: Proceedings of the 8th Conference on USENIX Security Symposium* (Vol. 8, pp. 14–14). Berkeley, CA: USENIX. https://dl.acm.org/doi/abs/10.5555/1251421.1251435

Wu, J., & Tran, N. (2018). Application of blockchain technology in sustainable energy systems: An overview. *Sustainability, 10*(9), 3067. Retrieved August 24, 2019, from http://www.mdpi.com/2071-1050/10/9/3067

Yoo, Y., Boland, R. J., Lyytinen, K., & Majchrzak, A. (2012). Organizing for innovation in the digitized world. *Organization Science, 23*(5), 1398–1408. https://doi.org/10.1287/orsc.1120.0771.

Zimmermann, H. (1980). OSI reference model - the ISO model of architecture for open systems interconnection. *IEEE Transactions on Communications, 28*(4), 425–432. https://doi.org/10.1109/TCOM.1980.1094702.

Chapter 5
Distributing and Democratizing Institutional Power Through Decentralization

Amir Fard Bahreini, John Collomosse, Marc-David L. Seidel, Maral Sotoudehnia, and Carson C. Woo

5.1 Introduction

While just over 10 years old, early estimations suggest blockchain will store over 10% of global GDP in the next 10 years (World Economic Forum 2015). Thus, there's no surprise that this technology is touted as the "next big thing," which can reshape the global economy and replace many of our current institutional infrastructures (Iansiti and Lakhani 2017; Seidel 2018). At its core, blockchain is a distributed database aimed at increasing the integrity of data by storage in immutable, secure, and transparent blocks (Lemieux 2017). Once a valid user adds to the database, the record cannot be altered or removed unless a pre-determined percentage of users agree to do so (creating immutability) (DuPont 2019). Further, the records are cryptographically hashed (ensuring security) and can be viewed by all individuals with access to the system (providing transparency). One key feature that differentiates blockchains from older databases and ledger technologies is their decentralized nature. Indeed, the decentralized attribute of a blockchain creates everything from excitement to concern across sectors, governments, institutions, industry, and end-users (Tapscott and Tapscott 2016). Despite the popularization of blockchain technology and discourses surrounding decentralization, we still need critical reflection about what decentralization in blockchain means (Walch 2019).

A. Fard Bahreini (✉) · M.-D. L. Seidel · C. C. Woo
Sauder School of Business, University of British Columbia, Vancouver, BC, Canada
e-mail: amir.fard@sauder.ubc.ca; seidel@mail.ubc.ca; carson.woo@sauder.ubc.ca

J. Collomosse
Centre for Vision Speech and Signal Processing (CVSSP), University of Surrey, Guildford, UK
e-mail: j.collomosse@surrey.ac.uk

M. Sotoudehnia
Department of Geography, University of Victoria, Victoria, BC, Canada
e-mail: msotou@uvic.ca

V. L. Lemieux, C. Feng (eds.), *Building Decentralized Trust*,
https://doi.org/10.1007/978-3-030-54414-0_5

Many blockchain studies rely on a singular technical viewpoint when considering decentralization. From this perspective, decentralization refers to the removal of the central node, thereby eliminating single points of failure. Consequently, much of the literature focuses mainly on the architecture of decentralization, its influence, and potential technical hurdles (e.g., scalability, security) (Zheng et al. 2017). In recent years we have seen an exponential growth of blockchain applications in financial services, Internet of Things (IoT), and supply chain management. As blockchain protocols and implementations continue to evolve, we need to reflect on what decentralization means and to whom. Indeed, definitions of the term "decentralization" continue to vary across disciplines, sectors, and communities, highlighting several gaps in the literature.

This definitional ambiguity is further confounded by discussions about blockchains that conflate decentralization with distribution. Rather than tracing out an essentialized and fixed definition of decentralization by asking whether decentralized and distributed systems are the same or distinct, we propose a holistic framework. Walch's (2019) recent critical investigation of blockchain decentralization recognizes the concept as a constellation of processes, or a set of relational interactions between diverse entities across three distinct layers: data, social, and technical. A data-social-technical framework exposes collision points between and across diverse (and divergent) perspectives and methodologies on blockchain decentralization, while also encouraging interdisciplinary explorations of productive tensions surrounding the term. A relational definition of decentralization, therefore, recognizes the porosity of the term's boundaries, which can alleviate current points of definitional contention while also yielding more context-specific examinations of different attempts to implement decentralization. Advocating for a more flexible definition of the term decentralization, we argue, fosters pluri-disciplinary collaboration as it encourages scholars across diverse fields to develop a definition we can jointly build upon.

We, therefore, argue that blockchain technology is more than digital technology. Rather, blockchains, like many other digital media, organize and impact the way we collect, commodify, share, and understand data while also shaping social relations in variegated ways and with different effects. For instance, from a technical standpoint, the primary impact of decentralization might be considered to be the removal of a single point of failure, or a data perspective may emphasize the archival benefits to propagating copies of data across nodes in a network. Meanwhile, organizational theorists have a long tradition of understanding how decentralized organizing can gain power over highly-institutionalized central actors (Yue et al. 2019), while also creating potential frictions in resolving cross-functional conflicts (Young-Hyman 2017). Similarly, powerful central actors can use decentralized mechanisms, such as public impression management, to garner additional power (Cole and Chandler 2019), while also being influenced through enhanced trust with less central actors (Haveman et al. 2017). Collaboration between central actors and subgroups of decentralized ones can create asymmetric power structures (Curchod et al. 2019).

The relationship between decentralization and power is a complex, socially embedded one. Technical choices in design directly impact social and data components and create the need to consider constructs such as governance when making technical choices (Schmeiss et al. 2019). Therefore, in the blockchain decentralization context we need to understand the type and degree of effects decentralization has at the data, social, and technical layers, and how these effects are transitive across layers, if at all, when considering the ultimate governance structures.

Finally, after sketching out a relational framework for decentralization, surfacing how different attempts at decentralization influence technical, social, and data layers, we address the potential hurdles in the adoption of decentralized technologies such as blockchain. Whenever an innovation can have such a significant influence on the democratization of institutional power, those in power can resist. These institutional barriers to adoption must be understood as part of the design of the technology to help realize its ultimate societal success. The decentralization aspects of the technology may themselves create barriers to adoption, unless properly designed and implemented.

5.2 Concept Ambiguity

5.2.1 The Spectrum of Centralization and Decentralization

As we start to define and delineate decentralization, we must first question the overly simplified polar definitions of centralized and decentralized. One way to think of decentralization is as a process of becoming, where decentralized systems undergo potential changes that may or may not actualize (Deleuze and Guattari 1983). Processes of decentralization may be established via technical means (e.g., through protocol design), but may change over time as people make decisions (e.g., the social layer) about what transactions should be sanctioned on the blockchain (e.g., the data layer) (Walch 2019). In the wake of the 2016 DAO exploit, for instance, a cadre of Ethereum developers made independent decisions impacting the network (DuPont 2019; Walch 2019). Initial reactions to the exploit were subsequently followed by a community-informed (but not universal) decision to roll back the blockchain, effectively undoing the exploit (DuPont 2019; Walch 2019). Such events reveal the complexity surrounding implementing a decentralized system, prompting us to question whether fully-decentralized systems actually exist. If they do, are they long-lasting or temporary states? And even if they do not, and there is a spectrum of decentralization, what does decentralization do? Who gets to participate in decentralized systems, and why does that matter?

Angela Walch (2019) argues decentralization is used primarily to describe diffuse power structures, and thus impacts all legal decisions surrounding blockchains. She goes on to conclude that we need more precision about which aspects of any complex system are in fact decentralized. Indeed, decentralization is often taken for granted as an inherent feature of a blockchain, with fledgling critical reflections surrounding what the term means, and to whom. As DuPont (2019) explains, definitional clarity within the blockchain literature persists due to the interchangeable usage of the terms “decentralized” and “distributed.” While Baran’s (1964, p. 1) classification of a decentralized network as one removing the need for “complete reliance upon a single point” (cf. DuPont 2019) has been popularized in discourses about blockchains and decentralization, DuPont (2019) rightly points out that, under Baran’s (1964) definition, decentralized networks are “common species of centralized networks, often organized hierarchically.”

Notably, centralization facilitates control and concentrated leadership. Decentralization often is seen as facilitating democracy and fairness. The ideal level of decentralization and the need to dynamically adapt depends on the circumstances of the decentralized solution. For example, when there are many disagreements among the stakeholders of a decentralized system, it can result in inactivity or a bifurcation of the community. At such times, some forms of centralization can help get some problems fixed, with an ultimate transition back to decentralization if there is decentralized buy-in to the solution. Such dynamic processes may ebb and flow as needed to keep the overall health of the decentralized community intact. Pockets of centralization can grow, solve issues, and disband in different locations over time while still maintaining a healthy overall decentralized community.

Decentralization may instantiate in a system over time, even if momentarily, or dissipate. If we think of decentralization as an ongoing process without an ideal end-state, then we can begin to trace out a pluri-disciplinary framework to consider decentralization in its varied (and shifting) instantiations. A process-based approach to decentralization exposes where, how, and when power operates through distributed systems. Tracing the power relations embedded in decentralized communities, practices, and systems foregrounds discussions about the definitional clarity (or lack thereof) surrounding the decentralization and/or distributed characteristics of a system. These are important considerations for researchers and practitioners interested in blockchain technology.

5.2.2 The Distinction Between Decentralization and Distribution

Is there a difference between decentralization and distribution? Or, can the two terms be used interchangeably? Does decentralization have more to do with power (or democracy or decision-making) at the social layer, and distribution more to do with data (or processing and storing of data) at the technical layer? If the distinction is between the social and technical layers, then how are they related? What about the data layer?

How is the structure of decentralization determined? In particular, how does such a structure help to achieve consensus? In the case of Bitcoin, the decentralized structure appears simple as all full nodes have the same authority in everything. In other applications (e.g., healthcare), not all stakeholders can have, or *should* have, the same authority in everything due to their specialized role or duty (e.g., what a hospital can do vs. what a pharmacy can do).

Decentralization offers one way to manage complexity at all three layers. It is not possible to develop all applications (data and technical layers) or to include all stakeholders (social layer) at once. Decentralization is a modular way to evolve, but modules and stakeholders need to work together after being included in the system. Decentralization also allows disagreements or multiple viewpoints (at the data and social layers) to co-exist without having to worry about how they can work together (e.g., different representations of data). Rules and policies, for instance, must be designed to facilitate the addition of new nodes to a permissioned blockchain. This is not simple as there can be issues of power, privacy concerns, and disagreements (on, for example, a consensus mechanism).

If the distribution is at the technical layer, then it is a performance issue. When a transaction contains a very small amount of data, then replicating all the same data in all the nodes is feasible. When a transaction contains a large amount of data or data from outside the blockchain, then there is the question of who replicates what data. The more nodes replicating the same data, the worse its performance will be. Perhaps the trade-off is at the data layer where integrity, authenticity, security, and trust need to be taken into consideration when trading off how much to replicate in the distribution. It is this interrelationship among the social, data, and technical layers that makes the distinction between decentralization and distribution confusing and hard to define.

5.3 Influence of Decentralization in Blockchain

To assess decentralization from the data-social-technical perspective, we propose a new theoretical framework encompassing the influences of decentralization in all three areas. From a *data* perspective, decentralization characteristics of blockchains can have significant consequences on its security, namely, the confidentiality, integrity, and authenticity of data. Additionally, decentralization can give more power to users in regard to controlling their data. From a *technical* standpoint, blockchain creates an immutable ledger that removes the possibility of a single point of failure and also removes the central authority, giving more powers to users along the way. The technical layer is at the heart of the "decentralization impact"; we contend that any changes that happen (and will happen) under the social and data layers are due to changes in technical layers. For that reason, the technical layer is heavily interconnected with the other two layers. Finally, from a *social* perspective, decentralization can change the trust dynamic between individuals and systems and modify the asymmetric power dynamic between users and monitoring central agencies.

In the following sections, we discuss these distinct yet inter-correlated consequences. We'll also discuss the role of power. The status quo power structures in society influence all three layers (i.e., data, social, technical) and the process of

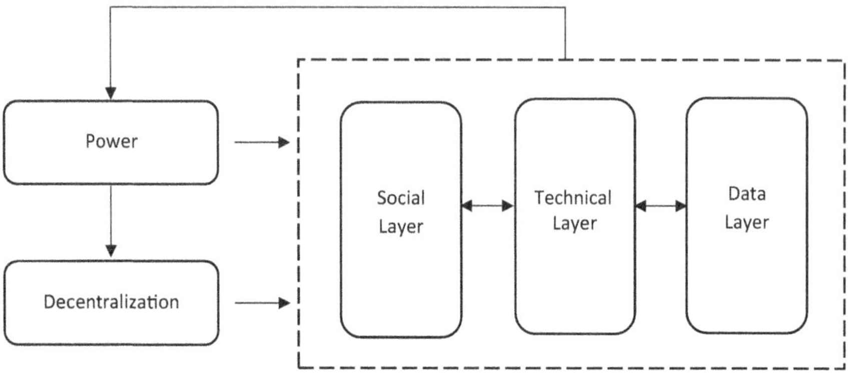

Fig. 5.1 Canonical model of the influence of decentralization from a data-social-technical framework

implementing decentralization. In other words, power in its various forms organizes everything and isn't merely a construct that is only influenced by decentralization. Instead, the current power holders have a direct influence on the process of decentralization at all three layers and may either encourage it or hinder it. This may get even more complex with them relinquishing one form of power (e.g., social) and strengthening another (e.g., technical). Thus, power appears to shape everything, and then those changes in the various layers will subsequently change existing power structures (see Fig. 5.1).

5.3.1 Influence on Data Layer: Improving Security and Disruptive Power of Data Decentralization

Over the past 10 years, people's control over their data has changed dramatically. With advances in machine learning and artificial intelligence (AI) over the past 5 years, deep learning—the use of deep neural networks (DNNs) to learn complex tasks—has transformed the landscape of how data is preserved and reused. DNNs are enabling a slew of new technologies that will transform our society—autonomous transportation (vehicles, drones), embodied agents for at-home care (avatars, robots), and autonomous decision making within financial services (e.g., risk underwriting)—all of which rely upon processing and analyzing users' data. Thus, there's a need to advance our analytical requirements while maintaining the confidentiality, integrity, and authenticity of data. Distributed Ledger Technologies (DLT), such as blockchain, offer such a solution via their immutable and decentralized infrastructure.

In this section, we first discuss how recent innovations in DNNs have changed both the way individuals' data is used, and how value is ascribed to it. Subsequently, we discuss how this usage has created numerous issues within conglomerates, such as Google and Facebook. Finally, we discuss how blockchain helps with the security of data, creating new forms of the digital economy that can address the power asymmetry between data owners/users and data subjects.

DNNs are a truly disruptive technology; their success at general-purpose learning tasks has driven their enthusiastic adoption across industry and academia. Yet DNNs require training, using vast quantities of computational power and data. A well-trained DNN has a significant intellectual property value due to this investment and its potential commercial impact. Since computational power can now be cheaply bought on the open market (cloud computing), the main enabler to producing lucrative DNN models is data, and lots of it. Tech giants (e.g., the FAANGS; a colloquial term for Facebook, Amazon, Apple, Netflix, Google, and similar organizations) have created a multi-billion dollar industry out of the collection and archiving of vast quantities of data collected over longitudinal time periods from individuals in exchange for digital services provided at no financial cost.

Yet society is rapidly coming to a realization that yielding data in this way has ceded power to tech giants in a way that does not accord with today's values. Recent legislation such as the European General Data Protection Regulation (GDPR) and high profile "data disasters" such as the Facebook/Cambridge Analytica scandal have raised public awareness of privacy and the value of data on the open market (Houser and Voss 2018). People have become more aware and cautious about sharing their personal data, uncomfortable with the disconnect between these billion-dollar industries, and the lack of compensation for their data used to build them. Within the past year, we have witnessed public outrage at the realization that personal data openly published on the internet (such as facial photographs) have been used to train DNN models.

The resulting media storm has caused some reputational damage to companies, and caused others to pre-emptively restrict access to large datasets they had made available, e.g., to academia for training DNNs. For example, Microsoft very recently removed their MSCeleb dataset of ten million faces amid accusations of racial and gender bias in the data distribution, and Facebook severely restricted programmatic access (the "API") to its platforms, preventing large-scale data mining of Facebook and completely restricting it for Instagram. These decisions to restrict access were made purely on commercial grounds and without consultation with academics and smaller businesses who rely upon that research data infrastructure. In other examples, commercial organizations were harvesting data (e.g., 3D models or facial photos) from the FAANGs' sites to train their DNNs, causing the FAANGs to act to restrict or retract their data and impacting the reproducibility of studies that had used that data, again eroding the research data infrastructure and stifling innovation.

More broadly, the digital platforms that underpin today's digital economy are driving a decentralization of opportunity and marketplace; everyone has the opportunity to be both a producer and consumer of goods and services through platforms like Uber, AirBnb, and YouTube. This creates an inexorable shift from classical economic models centred upon monolithic institutions, to a dynamic and decentralized peer-to-peer economy. But these dynamic, peer-to-peer markets are all underpinned by centralized digital platforms—run by FAANGs and other large software corporations. These companies make governance decisions unilaterally in isolation of the global impacts they have on societies. The realization is that whilst the marketplace for digital products, content and services has become decentralized—the power in the platforms underpinning this economy remains firmly with those who control the data—a position that lacks agency, and often results in decisions incompatible with the public good. Users lack agency over how their raw data is used, and corporations are increasingly reluctant to share it outside their boundaries. From a technical perspective, this has resulted in behavior that hinders innovations, particularly among small- and medium-sized (SME) enterprises, and led users to crave better control over their personal data whilst retaining access to the digital services modern society has come to rely on. From a societal perspective, the result is a centralized power structure derived from personal data that is not operated by those who contribute their data to it.

A natural reaction, particularly for the public, has been to change behavior; to become more selective or to desist in offering up their data for centralized siloing by tech corporations (e.g., the #deletefacebook movement). Users have pushed back, archiving data locally (on physical devices or within a private cloud data lake). Several research projects have explored platforms (e.g., the UK's Engineering and Physical Sciences Research Council (EPSRC) has funded projects Databox (EP/N028260/1) and Home Hub-of-all-Things (HAT) (EP/K039911/1)) for siloing social media or home IoT sensor data, whilst enabling sharing of data to corporations in return for some kind of compensation, e.g., money or services. Yet keeping data in a box and enabling users to make aperiodic, infrequent decisions to sell that data does not scale to release the large volumes of data necessary to train viable DNN models. Nor are there any guarantees available from personal data silos on the authenticity or provenance of data, so attributing value that might incentivize a corporation to enter into a data commodification economy. Therefore, there is a clear need for a way for users to increase security and retain control (agency) over their own data, which increases its value through greater data fluidity.

DLTs such as blockchain have the potential to offer such a solution, providing a means to ensure the integrity and provenance of personal data via hashes stored on-chain, but enabling users to retain the data off-chain. Such an infrastructure can also be used to broker access to data via smart contracts that exchange access (via encryption keys) for micropayments (Murray et al. 2019). In the UK, the EPSRC-funded Co-operative Models for Evidence-based Healthcare Redistribution (ComeHere) project (EP/P03196X/1) developed such a system to broker access to wearable fitness band data to healthcare insurance providers, e.g., AXA, with a vested interest in developing AI models of individual healthcare. In a traditional, centralized data economy the fitness band provider siloes raw data from individuals' fitness bands in return for providing personalized fitness services to the individual—yet that data is also exploited via sale to third parties out of the users' control. In the ComeHere model (Franceschi et al. 2018), the healthcare provider requests healthcare data, e.g., 2000 people's diabetes data. The request is brokered via a smart contract system which issues an offer to the individual. If the offer is accepted, the infrastructure provides those facets of the individual's wearable data to the healthcare provider in return for micropayments of cryptocurrency direct to them. The exchange is mediated through the exchange of encryption keys. As a result, the individual retains agency over what facets of their data are shared, and with whom, and they are directly compensated through the commodification features of the platform. This is just one emerging example of a new personal data revolution, in the spirit of personal data trusts (Hardinges 2018), enabled by a decentralized data sharing framework that enhances the privacy of the user whilst generating value for both them and the corporations that wish to build models from their data. The value ascribed to the data is due both to the provenance of the data underwritten by DLT, and the data fluidity enabled by the autonomous data brokering via smart contracts.

Another emerging form of data decentralization is federated machine learning, where multiple independent parties collaborate to train an AI model, while retaining decentralized control of their own data (Yang et al. 2019). Although federated machine learning has been researched for some time, that research has focused upon reducing the time taken to train models, for example, across multiple graphics processing units (i.e., GPUs, the hardware used to train DNNs) or a large corporate compute cluster or cloud. DLT is beginning to be explored for large-scale federated machine learning in untrusted scenarios, where each node in the DLT network is being operated by an independent entity that may not be trusted or might even act in an adversarial way (e.g., conducting a poisoning attack). In such situations, the DNN models may be trained collaboratively without parties necessarily needing to share their training data (which may be proprietary), whilst ensuring that no individual party can degrade or corrupt the model. Examples might include AI modelling for detecting pedestrians in an autonomous navigation system for cars—massive amounts of data are collected by manufacturers of autonomous cars, but this is siloed; no individual party shares their proprietary data, but all parties have a vested interest in collaborating to produce a pedestrian detection model that works with high accuracy.

Ultimately, these examples of data decentralization via DLT do more than enhance the security of data (i.e., confidentiality, integrity, and authenticity); they have given more power to users to control their data and only share what and when they deem necessary and beneficial.

5.3.2 Influence Technical Layer: The Disruptive Power of System Decentralization

From a technical standpoint, the decentralization of blockchain removes the need for validation of transactions from a trusted central node or agent. Over the past several years this characteristic has been highlighted by cryptocurrency communities wherein there have been discussions about elimination of the need for a central bank in financial transactions. One such example is Bitcoin, where parties can exchange currencies without the presence of a central monitoring agent, based on the proof-of-work consensus mechanism built in the system. While Bitcoin enables what many might consider decentralized finance, its very decentralization also serves as a non-trivial weakness to the system, making it difficult to scale the network effectively without compromising its decentralized nature. Further, determining which metrics to use when assessing Bitcoin's decentralization is in itself a tricky business. Gencer et al. (2018), for instance, explain how higher bandwidth allocation to node clustering and fairness variance (among other factors) can impact decentralization.

In addition to increasing users' responsibilities, a distributed trust model also influences the role of "accountability" throughout the structure. What if something unexpected happens? No system is without errors, and there will be times where an error can lead to a loss to users of the system. Under the current central structure, users not only trust the central node with the responsibility of verification but also with accountability. If an error occurs, users normally contact the central agent to seek answers, and even if the central node does not take responsibility for an error (as we have seen many of them don't), they at least must provide an answer. However, under the distributed structure, who will be responsible for responding to users and held accountable for errors?

5.3.3 *Influence Social Layer: The Disruptive Power of Transaction Decentralization*

While the answer to the potential consequences of blockchain in the data and technical layers is theoretically more apparent and predictable, blockchain's impact on society is still ambiguous. Most importantly, assuming that blockchain technology becomes pervasive in the next decade as some predict, how will this technology substitution (i.e., replacing existing centrally governed systems with a decentralized system) impact society, organizations, and people using the technology? While presenting the complete answer to this question is out of the scope of this chapter, we argue that two dynamics in society will drastically change: *power* and *trust*.

In the existing societal structure, most financial power resides with central agents (e.g., in financial transactions, this would be the banks). Under a decentralized system, this power will ideally diffuse among all system users, eliminating existing power asymmetries. However, the question becomes, can perfect diffusion of power occur under decentralized blockchain technology? To answer this question, we must look at trust and the outcome of decentralization from a socio-technical perspective. We present the example of Bitcoin to illustrate some of the ways decentralization of the network remains contested/is called into question.

Turning to the shift in trust dynamics and discuss its potential outcomes, trust has long been the focal point in the information systems literature as one of, if not, *the* most important antecedent of adoption and usage of new forms of technologies (Gefen et al. 2003; Hoffman et al. 2013). The blockchain ecosystem is often labeled as trustless, referring to the supposed obsolescence of verification from trusted third parties. Rather, with blockchain technology, it is consensus mechanisms that ensure the authenticity and integrity of the transactions. While blockchains appear to remove the need for trust in a central agent, users must still trust the system, and those running it, to use it. Theoretically, this results in a change from one-to-one relationships (e.g., users trust central agents) to one-to-many dynamics where users must be able to trust not only the system (i.e., blockchain), but also individual nodes.

5.3.3.1 Decentralization from a Socio-Technical Perspective

The importance of viewing decentralization from a socio-technical perspective is well illustrated by Bitcoin, currently the largest at-scale deployment of a permissionless blockchain infrastructure. Viewed purely from the technical dimension, the Bitcoin blockchain is a proof-of-work system. Bundles of financial transactions are appended to a timestamped ledger of previously vetted "blocks" of other financial transactions. Blocks are only considered valid on the chain if the hash of the transactions within them plus some padding (i.e., a "nonce") all hash together to produce a specific bit pattern. By design, it is computationally challenging to find a nonce that satisfies this rule—requiring massive computational or "mining" effort. The person who "mines" the block receives a financial reward in the form of bitcoins, incentivizing participation to maintain the network.

This arrangement appears fair, but given the odds of finding the correct nonce are infinitesimal, a social norm has emerged in which miners have teamed up, self-organizing into collectives called "pools", the largest of which are commercial organizations whose business is to win the race to mine the next block. At the time of writing, approximately 85% of Bitcoin blocks are mined by just ten mining pools commanding the majority of computational "hash power" on the blockchain (Blockchain Luxembourg SSA n.d.). The social dynamics of Bitcoin's blockchain are thus unfair and centralize power in the hands of a coterie of miners. This, in turn, makes it possible for pools to collude to fork the blockchain or to manipulate the virtual currency market: a distribution of power that is far from decentralized.

Consensus protocols are diverse, and many alternatives to proof-of-work chains exist. All appear at face value to offer an even technical playing ground to network participants, yet the social dynamics discussed above do not necessarily result in decentralization. Proof of authority networks (a common choice for permissioned blockchains) require at least half of the nodes active on a blockchain to seal (i.e., agree on) a new block before it is admitted to the chain, or to perform administrative actions such as admitting a new node onto the permissioned chain. But many factors can corrupt this system, which fundamentally assumes independently organized nodes. Collusion and faction building between nodes can influence willingness to sign, and even security issues come to bear—were a denial of service (DoS) attack mounted upon sufficient nodes to prevent their interaction with the network, the remainder of nodes, if colluding, could agree to admit a disingenuous block to the chain—a so-called "51% attack".

This underscores the importance of considering decentralization and institutional power within blockchain-mediated systems from both the social and technical dimensions. Thus, it appears that there is a possibility that, instead of even distribution of power among users, groups of users form to gain power within the system. While diffusion of power on the socio-technical level is an ideal outcome of a decentralized blockchain implementation, the distribution of power risks becoming compromised by collusion on the network.

5.3.3.2 Decentralization and Distributed Trust

One of the ideal outcomes of blockchain decentralization involves removing the need for a trusted third party to verify transactions (e.g., central agents such as banks for financial transactions). While this may be technically possible on a blockchain, trust is nevertheless downloaded onto individual users of the system. As a result, greater responsibility is placed on the individual to maintain the integrity of the network.

There seems to be one major outcome under the decentralized structure from the user perspective: an increase in responsibility. Under this distributed form of trust, every user is relying upon others' work to assess the authenticity and integrity of transactions. Simply put, all users are responsible for verifying transactions based on the data available in public ledgers. Therefore, even though blockchain is commonly referred to as the "trustless" machine, the word trustless refers to the central node. This is because, in reality, the notion of trust doesn't become irrelevant under this technology, but rather reshapes to "distributed trust," which subsequently leads users to take up more responsibilities.

5.3.4 Resistance to the Decentralization Process

Central positions in the societal field are traditionally thought of as a source of power (Seidel 2017). Transitioning to a decentralized organization of activity significantly reduces such centralized power benefits, reducing the competitive advantages of defending such positions, and thus potentially shifting societal power. The process can be held back directly by the maintenance and ongoing dominance of powerful institutions such as corporations, governments, and religious organizations (Yue et al. 2019).

Shifting to such decentralized models is a difficult process, laden with dominant power structures holding things more centralized, or in pockets of recentralization at the very least. These types of centralizing inertia or recentralization power are further bolstered by the financial resources typically attributed to dominant central power positions (Chen and Bellavitis 2020). This creates a fundamental challenge for the decentralized model and implies the need to conceive of it as a transitionary process instead of a spectrum or state.

Democracy is a political ideal that frequently has hidden pockets of institutional power embedded both explicitly and implicitly. Centralized powerful actors can co-opt democratic systems, and shift outcomes to those that benefit their own self-interest. True decentralization of power parallels the existential challenges of true democratic systems and is subject to the same co-optation pressures. When considering decentralization processes and technologies, we therefore also need to give consideration to the legal and governance challenges (Halaburda et al. 2019).

The design of the technical, data, and social elements of decentralized systems is key to the ultimate success of the decentralization of power. It offers opportunities for the structurally less powerful of society to reclaim power over democratic processes if the core technical dimensions are designed with such true democratic ideals in front of mind.

5.4 Conclusion

In this chapter, we attempted to address three gaps in confounding debates about the "decentralized" attributes of blockchains: ambiguity, decentralizing effects, and some of the limits to decentralization (including widespread adoption). Decentralization, we argued, involves a process of actualization, or becoming, whereby the diffusion of control and coordination over a network of peers and/or power in a system are not necessarily inherent attributes, but involves what Deleuze and Guattari (1983, p. 19) might call "bands of intensity, potentials, thresholds, and gradients". Our motivation was, therefore, to present a concise yet holistic description of decentralized characteristics of a blockchain to understand better how decentralization operates in diverse contexts and across different data, social, and technical registers. In this process, we drew upon prior literature and distinguished between decentralization and distribution, and we proposed a relational conceptual framework for blockchain decentralization. A relational definition of decentralization acknowledges the ambiguity of the term, while also recognizing that decentralization is not merely an antipole of centralization, but rather a continuum with many levels, changing based on different contexts, and rife with contradictions.

We next considered the effects of practices of decentralization across three layers—the data, social, and technical—to better understand how decentralization takes shape, or becomes, and where it might fail. While blockchains enable diffusion of power across networks at every layer, for instance, attempts to decentralize using blockchains do not necessarily result in greater network resiliency, greater network access, participation, or equity. Additionally, while blockchains encourage decentralization of data and can empower users to regain agency over personal data otherwise commodified by large technology firms such as Facebook or Google, it remains complicated and computationally intensive.

Given the relative infancy of blockchains, the remarks in this chapter are prefatory. In addition to the technical advancement of decentralized systems, we have discussed the disruptive power of data decentralization and existing challenges in relation to data preservation and security. As the process of decentralization evolves, the need for checks and balances, and enhancements in data-use and data-privacy policies, will increase and require further work. From a societal perspective, we must be aware of the resistance and pitfalls in the democratization process. Whenever the possibility of diffusion of power emerges, resistance emerges that may defeat the purpose of such change. The Roman Republic was supposed to be a democracy but as history reads, that promise was never fulfilled and power was distributed between only a group of families, forming an oligopoly. Nowadays, a few tech conglomerates hold considerable centralized power over users' data, despite the promise of distributed power. The process of decentralization still has a long way to go, and social scientists should play a large role in shaping its progress in the future. With that said, further research on the decentralized attributes of blockchains could examine context-specific implementations of decentralized protocols to better understand how decentralization becomes possible, and trace out some of its empirical and material realities and contradictions. Doing so would open up further research trajectories surrounding the data, social, and technical impacts of blockchains, and expose some of the potential challenges with maintaining decentralized systems in an effort to truly democratize institutional power.

References

Baran, P. (1964). *On distributed communications: I. Introduction to distributed communications networks*. Santa Monica, CA: RAND Corporation. Retrieved from https://www.rand.org/pubs/research_memoranda/RM3420.html

Blockchain Luxembourg SSA. (n.d.). *Hashrate distribution*. Retrieved from https://www.blockchain.com/en/pools

Chen, Y., & Bellavitis, C. (2020). Blockchain disruption and decentralized finance: The rise of decentralized business models. *Journal of Business Venturing Insights, 13*, e00151. https://doi.org/10.1016/j.jbvi.2019.e00151.

Cole, B. M., & Chandler, D. (2019). A model of competitive impression management: Edison versus Westinghouse in the war of the currents. *Administrative Science Quarterly, 64*(4), 1020–1063. https://doi.org/10.1177/0001839218821439.

Curchod, C., Patriotta, G., Cohen, L., & Neysen, N. (2019). Working for an algorithm: Power asymmetries and agency in online work settings. *Administrative Science Quarterly.*, 000183921986702. https://doi.org/10.1177/0001839219867024.

Deleuze, G., & Guattari, F. (1983). *Anti-Oedipus: Capitalism and schizophrenia*. Minneapolis, MN: University of Minnesota Press.

DuPont, Q. (2019). *Cryptocurrencies and blockchains*. Cambridge: Polity Press.

Franceschi, M., Morelli, D., Plans, D., Brown, A., Collomosse, J., Coutts, L., & Ricci, L. (2018). ComeHere: Exploiting Ethereum for secure sharing of health-care data. In G. Mencagli et al. (Eds.), *Euro-Par 2018: Parallel processing workshops – Revised selected papers, Euro-Par 2018. Lecture notes in computer science* (Vol. 1133, pp. 585–596). Cham: Springer. https://doi.org/10.1007/978-3-030-10549-5_46.

Gefen, D., Karahanna, E., & Straub, D. W. (2003). Trust in TAM in online shopping: An integrated model. *MIS Quarterly, 27*(1), 51–90.

Gencer, A. E., Basu, S., Eyal, I., Van Renesse, R., & Sirer, E. G. (2018). Decentralization in bitcoin and ethereum networks. In S. Meiklejohn & K. Sako (Eds.), *Financial cryptography and data security (FC 2018) – Revised selected papers. Lecture notes in computer science* (Vol. 10957, pp. 439–457). Cham: Springer. https://doi.org/10.1007/978-3-662-58387-6_24.

Halaburda, H., Levina, N., & Semi, M. (2019). Understanding smart contracts as a new option in transaction cost economics. In *Proceedings of the 40th International Conference on Information Systems,* Munich. Available at SSRN: https://ssrn.com/abstract=3506223

Hardinges, J. (2018). *Defining a 'Data trust'*. Open Data Institute. Retrieved from https://theodi.org/article/defining-a-data-trust/

Haveman, H. A., Jia, N., Shi, J., & Wang, Y. (2017). The dynamics of political embeddedness in China. *Administrative Science Quarterly, 62*(1), 67–104. https://doi.org/10.1177/0001839216657311.

Hoffman, R. R., Johnson, M., Bradshaw, J. M., & Underbrink, A. (2013). Trust in automation. *IEEE Intelligent Systems, 28*(1), 84–88. https://doi.ieeecomputersociety.org/10.1109/MIS.2013.24

Houser, K. A., & Voss, W. G. (2018). GDPR: The end of Google and Facebook or a new paradigm in data privacy. *Richmond Journal of Law & Technology, 25*(1). https://jolt.richmond.edu/gdpr-the-end-of-google-and-facebook-or-a-new-paradigm-in-data-privacy/

Iansiti, M., & Lakhani, K. R. (2017). The truth about blockchain. *Harvard Business Review, 95*(1), 118–127. https://hbr.org/2017/01/the-truth-about-blockchain

Lemieux, V. L. (2017). A typology of blockchain recordkeeping solutions and some reflections on their implications for the future of archival preservation. [Paper presentation]. *2017 IEEE International Conference on Big Data (Big Data)*, Boston, MA, 2017 (pp. 2271–2278). https://doi.org/10.1109/BigData.2017.8258180

Murray, A., Kuban, S., Josefy, M., & Anderson, J. (2019). Contracting in the smart era: The implications of blockchain and decentralized autonomous organizations for contracting and corporate governance. *Academy of Management Perspectives.* https://doi.org/10.5465/amp.2018.0066

Schmeiss, J., Hoelzle, K., & Tech, R. P. G. (2019). Designing governance mechanisms in platform ecosystems: Addressing the paradox of openness through blockchain technology. *California Management Review, 62*(1), 121–143. https://doi.org/10.1177/0008125619883618.

Seidel, M.-D. L. (2017). Network opportunity emergence and identification. In M.-D. L. Seidel & H. R. Greve (Eds.), *Emergence: Research in the sociology of organizations* (Vol. 50, pp. 141–168). Bingley: Emerald. https://doi.org/10.1108/S0733-558X20170000050005.

Seidel, M.-D. L. (2018). Questioning centralized organizations in a time of distributed trust. *Journal of Management Inquiry, 27*(1), 40–44. https://doi.org/10.1177/1056492617734942.

Tapscott, D., & Tapscott, A. (2016). *Blockchain revolution: How the technology behind Bitcoin is changing money, business, and the world.* New York: Portfolio/Penguin.

Walch, A. (2019). Deconstructing 'decentralization': Exploring the core claim of crypto systems. In C. Brummer (Ed.), *Cryptoassets: Legal, regulatory, and monetary perspectives* (pp. 39–68). New York: Oxford University Press.

World Economic Forum. (2015). *Deep shift: Technology tipping points and societal impact.* Retrieved from http://www3.weforum.org/docs/WEF_GAC15_Technological_Tipping_Points_report_2015.pdf

Yang, Q., Liu, Y., Chen, T., & Tong, Y. (2019). Federated machine learning: Concept and applications. *ACM Transactions on Intelligent Systems and Technology (TIST), 10*(2), 1–19. https://doi.org/10.1145/3298981.

Young-Hyman, T. (2017). Cooperating without co-laboring: How formal organizational power moderates cross-functional interaction in project teams. *Administrative Science Quarterly, 62*(1), 179–214. https://doi.org/10.1177/0001839216655090.

Yue, L. Q., Wang, J., & Yang, B. (2019). Contesting commercialization: Political influence, responsive authoritarianism, and cultural resistance. *Administrative Science Quarterly, 64*(2), 435–465. https://doi.org/10.1177/0001839218770456.

Zheng, Z., Xie, S., Dai, H., Chen, X., & Wang, H. (2017). *An overview of blockchain technology: Architecture, consensus, and future trends.* [Paper presentation]. 2017 IEEE International Congress on Big Data (BigData Congress), Honolulu, HI, 2017 (pp. 557–564). https://doi.org/10.1109/BigDataCongress.2017.85.

Chapter 6
Blockchains and Provenance: How a Technical System for Tracing Origins, Ownership and Authenticity Can Transform Social Trust

Danielle Batista, Henry Kim, Victoria L. Lemieux, Hrvoje Stancic, and Chandana Unnithan

6.1 Introduction

The modern digital world has brought us new models of communication with both advantages and disadvantages. The World Wide Web has made it easier for us to produce and disseminate information and lowered barriers of communication to improve access to information. However, the same advances have raised serious concerns about trust in information. Blockchain technology is viewed as having great promise to address these concerns: This is why blockchain is also referred to as a "trust machine". But is that really the case? Can blockchain technology restore trust?

Trust is a very intricate term with different meanings and significance in different contexts. Nevertheless, there is some agreement with the idea that trust relies upon knowledge of the origin of something or someone. That is why provenance and trust are seen as being closely related concepts. For this reason, there is growing interest across many fields in the concept of provenance and its application as a vehicle to promote trust. Lemieux (2016) asserts that in computer science there has been

D. Batista (✉) · V. L. Lemieux
School of Information, University of British Columbia, Vancouver, BC, Canada
e-mail: danielle.batista@ubc.ca; v.lemieux@ubc.ca

H. Kim
Schulich School of Business, York University, Toronto, ON, Canada
e-mail: hkim@schulich.yorku.ca

H. Stancic
Department of Information and Communication Sciences, University of Zagreb, Zagreb, Croatia
e-mail: hrvoje.stancic@zg.t-com.hr

C. Unnithan
Department of Public Health, Torrens University Australia, Melbourne, VIC, Australia
e-mail: Chandana.unnithan@laureate.edu.au

V. L. Lemieux, C. Feng (eds.), *Building Decentralized Trust*,
https://doi.org/10.1007/978-3-030-54414-0_6

growing recognition of the necessity of developing applications to trace and analyze the provenance of data across increasingly distributed and networked computing environments. Yet, even though technical systems can be used to trace provenance, providing knowledge of the origins of something or someone in this way is only one dimension of the complex concept of trust. Thus, is it really possible that an emerging technical system, a distributed ledger, such as blockchain technology, can be used to solve a problem that has many dimensions and is also social in nature? By exploring the use of blockchain technology to trace provenance, and its implications for social trust, our aim in this chapter will be to offer a tentative answer to this question.

The chapter begins with a brief overview of different conceptualizations of provenance and recent applications of it in the fields of health and business. To structure our exploration of blockchains, provenance and social trust, we rely upon a broad framework previously used in the analysis of the trustworthiness of ledger records in blockchain systems, the "taxonomy of trust" (Lemieux 2017), which derives from the discipline of archival science (Fig. 6.1). The taxonomy presents

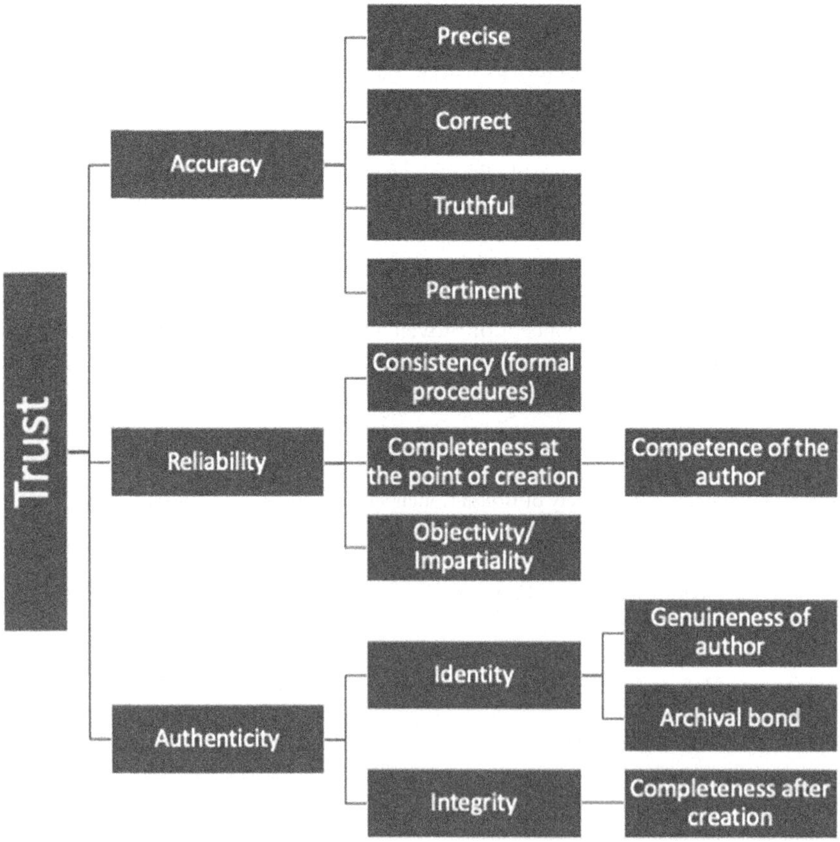

Fig. 6.1 A taxonomy of key archival concepts and their relationship to trust (Reproduced from Fig. 9 in Hofman et al. 2018)

three major requirements for trustworthy records—accuracy, reliability, and authenticity—noting that continued reliance upon records also requires *determination* of whether the origins of records suggests that they are accurate and reliable and *preservation* of these characteristics in records over time, i.e., authenticity, or the ability to prove that a record is what it purports to be.

Each of these requirements will be explored in the next sections of this chapter. Section 6.3 discusses the potential to use token economics in real-life, i.e., cryptocurrency and incentives design, to improve the accuracy and reliability of data and records. The focus in this section is on the use of blockchain as a viable alternative to centralized systems, framing the decentralized versus centralized ledger dichotomy. In Sect. 6.4, the focus turns to blockchain and authenticity of digital records with a discussion of the concept of blockchain-based digital originals as a means to differentiate authentic originals from copies in the digital environment. Section 6.5 discusses the preservation of provenance, focusing on the relevance of provenance for long-term access to information in the digital environment.

6.2 Conceptualizations and Recent Applications of Provenance

The general concept of provenance provided by the OED Online is "[t]he fact of coming from some particular source or quarter; origin, derivation" and in relation to the notion established by the arts field "[t]he history of the ownership of a work of art or an antique, used as a guide to authenticity or quality; a documented record of this" ("Provenance," n.d.). These general concepts provide an overview of diverse conceptualizations of provenance across different cognate disciplines and fields of practice. For example, according to the International Council on Archives, provenance is "[t]he relationship between records and the organizations or individuals that created, accumulated and/or maintained and used them in the conduct of personal or corporate activity" ("Multilingual Archival Terminology Database," n.d.), while the Glossary of Archival and Records Terminology defines provenance as "1. The origin or source of something.—2. Information regarding the origins, custody, and ownership of an item or collection" Pearce-Moses 2005). The Encyclopedia of Database Systems defines data provenance as "[referring] to a record trail that accounts for the origin of a piece of data (in a database, document or repository) together with an explanation of how and why it got to the present place" (Gupta 2009). Despite the differences, there remain strong similarities in the conceptualization of provenance across disciplines and fields converging on the ideas of origin, ownership, and authenticity.

Importantly, we note that tracing and documenting provenance has been used to explore how to mitigate problems related to data and records that negatively impact upon the social sphere of trust, such as addressing recent concerns about the genuineness of data and records in the context of combatting "fake news". As another example, some studies of healthcare data provenance highlight that to provide better "user-centered healthcare services, the treatment of a patient requires viewing the processes and data as a whole" (Kifor et al. 2006). Xu et al. (2018) point to the relevance of data provenance for healthcare since it "maintains the integrity of the digital objects, e.g., the results of data analysis engender greater trust if their provenance shows how they were obtained" and affirms that "[i]n health data settings, [provenance] can be used to deliver auditability and transparency, and to achieve trust in a software system".

In the business context, notions of provenance are frequently linked to supply chain management. Globalization and the multiple company character of contemporary business models, followed by numerous scandals involving large corporations (e.g., use of slaves in manufacturing, lack of sustainability and suspicious quality of products delivered to consumers), has increased concern about the provenance of goods such as food and clothes. Blockchain platforms, which can be used to control the production of assets from raw material to final product, have been proposed as a potential solution to address these issues. As Kim and Laskowski (2018) assert:

> [e]valuating knowledge provenance has become more possible as more and more of the data required to discover the source of knowledge is recorded on the Web. Evaluating provenance of physical goods—or what we call supply chain provenance—has generally been more difficult because so many goods are handled in complex, international supply chains where granular tracking of physical characteristics and product whereabouts has not been possible. That is, until recently, when provenance evaluation has become more possible with the advent of IoT [the Internet of Things] and blockchain.

Thus, the trustworthiness of transaction records, whether in healthcare or business supply chains, and the trustworthiness of the data they convey, connect the technical aspects of blockchain systems with transformations in trust among social actors. We therefore argue that *tracing and documenting the provenance of data and records* are at the nexus of how blockchains can be used to *transform social trust.*

6.3 Token Economics in Real-Life: Cryptocurrency and Incentives Design to Improve the Accuracy and Reliability of Data and Records

In the taxonomy of trust model, accuracy and reliability are key to realizing the trustworthiness of data and records. Accurate records are precise, correct, truthful, and pertinent to the matter. Reliable records are characterized as consistent, complete, and objective. Accuracy and reliability are also referenced in the management information systems literature, albeit with slightly different connotations from their conceptualization in archival science, and they are associated with a wider range of data quality characteristics (Wang et al. 1995). These two characteristics of trustworthy records are often not found in centralized data and records stores. This may be the case because an intermediary has used its privileged position to exploit, rather than serve, the parties who entrust it with the privilege of storing their data and records. As a distributed ledger, blockchain's decentralized model of storing data across various stakeholders is meant to prevent this type of abuse of privilege. For blockchain to be a viable alternative to centralized ledgers then, the accuracy and reliability of the data and records held on the ledgers must be expected to be better than, comparable to, or at least not substantially inferior to the quality of the data and records held centrally. Thus, in this section, we frame the decentralized versus centralized ledger dichotomy associated with blockchain use as a data quality challenge, proposing that blockchain should be used when the quality of data and records that can be achieved on the blockchain is better than the quality of data and records stored by more centralized means.

The data quality challenge is a very nuanced one given that there are many dimensions of what constitutes quality in connection with data. A classic survey paper lists 26 dimensions (Wang et al. 1995). Of the 26, let us ignore dimensions such as relevance, content, and importance, which are more external to the design of a system. That is, whether the data are stored on a blockchain or a centralized database does not necessarily make the data any more or less relevant and important (roughly equivalent to the notion of pertinence in the archival science taxonomy of trust). What remains are characteristics that are concerned with an "internal view" of a system's design and operations (Wand and Wang 1996): Accuracy, reliability, timeliness, completeness, currency, consistency, and precision. This grouping of system-internal characteristics from Wang et al.'s (1995) data quality model is roughly equivalent to those encompassed within notions of accuracy and reliability in the archival science-based taxonomy of trust model. The Bitcoin network is the key exemplar for demonstrating these qualities. The *accuracy* of data on the Bitcoin network has been remarkably high. And the system, given that it is an open source project, is surprisingly reliable, leading then to *reliable* data. The data are *timely* insofar as they update system-wide every 10 min or so when a new block is added. The data on the network's blockchain are remarkably *complete* in that all Bitcoin transactions ever recorded are publicly accessible. A real-time record of unverified transactions is also available, providing for data *currency*. At each block creation, system-wide data *consistency* is achieved. And all network numerical data are generated in floating point, indicating high levels of *precision*.

However, the high data quality inherent in Bitcoin can be juxtaposed with counterfactual examples. Less popular cryptocurrency blockchains at times have been susceptible to "51% attacks" (Nayak et al. 2016) meaning that in those blockchains data and records are more likely to be inconsistent and even inaccurate. And outside of cryptocurrencies, "garbage in, garbage out" applies to blockchains as to any other system. That is, a very good, say, food tracing blockchain can ensure that inaccurate information stored on the blockchain will immutably remain inaccurate. Arguably, then, poor quality data recorded on the blockchain can jeopardize the usefulness of blockchain systems as much as poor quality blockchain system design. The example of the Bitcoin network points to the potential of blockchain technology to assure the accuracy, reliability, timeliness, completeness, currency, consistency, and precision of data. While this capability cannot be generalized to every blockchain system, as we have already pointed out, we wish to highlight that the application of blockchain and provenance tracking as a research area is addressable by referring to a large amount of research into data quality from the field of management information systems in combination with theories of records trustworthiness from the discipline of archival science.

In a blockchain operating for a supply chain of international partners over many tiers, the blockchain data would be sourced from a variety of different systems—running the gamut of sophistication from Enterprise Resource Planning (ERP) systems, Internet of Things (IoT) devices, to mere self-reporting—across different international jurisdictions. We see the possibility of using blockchain technology to ensure high quality input data despite the heterogeneity of input regimes. For instance, a consensus mechanism could operate over heterogeneous systems, automated and manual, to agree that a given piece of data to be input into a blockchain is accurate. Or, leverage the transparency, immutability, and auditability of blockchain to enforce practices from supply chain partners that will lead to higher quality data input. For instance, a blockchain solution could incorporate auditing and comparison of ledger records with physical measurements, so that cheats who mislabel can be readily caught and punished. In proposing a blockchain research agenda that focuses on the issue of data quality, it could be interesting:

- to frame blockchain justification or selection as a data quality problem; and
- to investigate data quality within the blockchain and also quality of data input into the blockchain.

To apply a blockchain metaphor to the problem of input data quality means, as a start, investigating: (a) consensus mechanisms operating over heterogeneous systems that provide data inputs into a blockchain, and (b) decision-making policies that use aggregated data on the blockchain to incentivize high quality data input from stakeholders. At this point in time, this remains as a promising open research agenda and the subject of possible future work, unlike the research discussed in the next section that is much further along.

6.4 Blockchain and the Authenticity of Digital Records: The Concept of Digital Originals

Records are commonly generated and found in recordkeeping and archival institutions and systems in both analogue and digital form. They are assessed according to the same foundational principles of archival science. The trustworthy records are assessed as accurate, reliable and authentic. Authentic records preserve their identity and integrity over the period of long-term preservation (Fig. 6.1). However, although analogue and digital records are assessed according to the same archival principles, the digital medium introduces volatility. While the content of the analogue records is dependent on the medium, digital content can be freely transferred from one medium to another (ideally) without any loss. Also, one can make as many copies indistinguishable from the original as one wants. The InterPARES Project (2001) differentiates between three types of copies: copy in the form of an original, imitative copy and simple copy, noting that:

> The most reliable copy is a copy in the form of an original, which is identical to the original although generated subsequently. An imitative copy is a copy that reproduces both the content and form of the record, but in such a way that it is always possible to tell the copy from the original. A simple copy is a copy that only reproduces the content of the original.

While copies can be reliably identified in the analogue form, it may not be that easy to do so in the digital form. Namely, if one makes a copy of a copy in the analogue form, the last copy will in most cases be easily detectable as clearly different from the original (e.g., because there will be tiny, but detectable variations in the production process as well as the possibility of detectable variances between the original and its copy of degradations in physical form). On the other hand, every copy of a digital record appears to be the same and it is, moreover, easy to create as many copies as one wants and proliferate them throughout a network. This brings us closer to the challenge examined here—management of digital originals and the deleterious effect on social trust of uncertain authenticity.

In order to better understand the complexity of the concept of "digital original", the term "original" should be defined. Pearce-Moses (2005) defines original as follows:

> n. ~ 1. The initial manifestation of something.—2. A thing from which copies are made, especially a prototype.—3. Diplomatics · The first complete and effective version of a record.—4. Law · The thing itself or a duplicate intended to have the same effect by the person creating it.

From this, it is clear that one has to be able to determine which of the two digital documents is the initial one in order to identify the original. It is easy if the two documents are different and both have time of creation indicated. However, it may be the case that an "original" and a "copy" appear the same, i.e., one could be dealing with a copy in the form of original (one of the two files is a copy of the other) in which case the two documents are indistinguishable. This means that it is possible to (theoretically) create an infinite number of originals. While this usually does not present any problem as long as the original that is in the possession of a particular person produces the intended effect or is a record of the decision acted upon, in certain cases the existence of an uncontrolled number of what appear to be originals poses a challenge. Negotiable instruments represent one such case.

The Business Directory (2019) defines negotiable instruments as:

> unconditional orders or promise[s] to pay, and include checks, drafts, bearer bonds, some certificates of deposit, promissory notes, and bank notes (currency). A negotiable instrument has three principal attributes: (1) an asset or property (that is the subject matter of the instrument) passes from the transferor to the transferee by mere delivery and/or endorsement of the instrument, (2) a transferee accepting the instrument in good faith and for value (and who has no notice of any defect in the title of the transferor) obtains an indefeasible title and

may sue on the instrument in his or her name, and (3) [that] no notice of the transfer need [s] to be given to the party liable in the instrument.

In the case of an analogue, paper-based negotiable instrument, it is clear that the one who possesses it is entitled to the rights arising from it, and that they, upon transferring it to another, no longer possess the paper negotiable instrument nor any rights arising from it. Therefore, possession of the paper original means control of the rights. This is not easy to accomplish in the digital form since numerous originals may easily be created. How can the original bearing rights be identified among many identical instances? Who has the "digital original" and who has the "copy"? Who has the right to sell? Or, who has the right to collect repayments?

The above-noted challenges give rise to several questions: Can we have only one digital original? Can we know who has it? Can we allow creation of an infinite number of copies, while still being able to distinguish, manage and control the original? Can we allow the transfer, buying and selling of the original and yet allow the initial owner to keep his or her copy (which was original before the transfer), i.e., can we allow transformation of an original into a copy? And finally, can we allow rights arising from the digital original upon transfer of possession? Although all this might seem impossible due to the nature of digital media, we offer the example of a blockchain-based system for management of digital originals of negotiable instruments as proof that it is possible to positively answer the above questions.

Using blockchain technology, digital originals can be realized as smart contracts or as Ricardian contracts. While smart contracts do not include semantics and some computer scientists might recognize them as a state machine, Ricardian contracts—a method of recording a document as a contract at law, and linking it securely to other systems, such as accounting, for the contract as an issuance of value (Grigg 2004)—allow richness of semantics, are structured and both computer and human readable. Most importantly, Ricardian contracts can be realised to mimic the "look and feel" of paper contracts, to describe content and rules of agreement, the intentions of the contract, and to match legal regulations. Figure 6.2 shows an example of a promissory note realised as a Ricardian contract.

The example in Fig. 6.2 shows how blockchain and distributed ledger technologies, which are based on four underlying principles—calculation of hash values, Merkle tree, chaining of root hashes, and distributed consensus—can rise to the challenge of creation and management of digital originals (see the Content/Technical Details section—lines 64–70—in Fig. 6.2). The system manages the creation of a single original of a promissory note by creating a cryptographic envelope around the initial content (i.e., its hash) and ownership information (i.e., the owner's private key), adding information about the version and time of creation (i.e., its timestamp). Then, the package is registered in the blockchain. Thus, the content is fixed, the owner is uniquely identifiable, and the time of creation can be confirmed. Multiple copies can be made but only the owner, using his private key, has the rights coming out of the promissory note, and only the owner can transfer it to another party or invalidate it. When the transfer is made, the amendment cryptographic envelope is

```yaml
###################################################################################
################ THIS IS A TEST DOCUMENT - NO CONTENT IS LEGALLY BINDING. #########
###################################################################################
# By signing the content in this digital original document and amendments hereto,
# using a digital signature, the signee accepts this digital original as the valid
# and legal bearer of it's content. This document is a versioned digital original
# secured by the Enigio trace:original system.
# Since the document is registered in the trace:original public ledger, nothing
# may be altered or removed without this file losing it's authenticity.
# All content is in YAML-format, cryptographically secured both in this file and
# by immutable references in the block-chained trace:original public ledger.
# Only those in possession of this trace:original file, or a copy thereof, have
# access to its business content.
# The authenticity of a trace:original file can be verified online as well as
# it's consistency off-line, by using the appropriate software and algorithms as
# found at https://traceoriginal.com.
# Only the holder of the current and valid trace:original file, together with it's
# current private key, is the party able to exercise the legal rights as described
# in the document. Furthermore, this current legal holder has the control of the
# document with the right to make amendments, facilitate the procedure of adding
# signatures, transfer ownership and invalidate the current digital original.
# If not stated otherwise, this digital document should be subject to the eIDAS
# regulation (EU) No 910/2014.
# Additional information can be found on https://traceoriginal.com.
###################################################################################
Version: 1.0
Content:
  Company credit: Promissory note 2019-1, Company Ltd, 551100ABC999ABCAB500
  Agreement no: 12345
  Debtor:
    Company: DebtorCompany Ltd
    LEI: 543200ABC111ABCAB111
    Adress: Drottningholmsvägen 10, 11242, Stockholm
  Contract period, Terms of Payment, Collaterals etc.:
    Credit Amount GBP (in letters): Ten thousand pound
    Credit Amount GBP (in numbers): 10000
    Credit period number of months: 6
    Due date repayment: 15th every month
    Amortization to pay each due date (GBP): 1666.66
    First due date amortization (year-month-day): 2019-05-15
    First due date interest rate(year-month-day): 2019-05-15
    Interest Rate % (Yearly): 8.00
    Due interest rate: 15th each month
    Payment term interest (monthly or quarterly): monthly
    Number of repayments: 6
    Setup fee (GBP): 0.00
    Notification fee (GBP): 2.90
  Terms and Conditions: |
    By signing the promissory note, the Debtor also approves Company Ltd General Terms (Appendix 1 referenced below)
    and shall pay to the Company Ltd, or order, the credit amount together with interest, fees and costs according
    to the provisions of this debt, some of which are in the General terms and condition for the credit. This agreement
    has been drawn up in an original from which the borrower has received a copy whose compliance with the original
    can be verified as described on the website https://traceoriginal.com
    This agreement is valid from the signing of the agreement.
  Attachment 1:
    Name: General Terms and Conditions 2019-1.pdf
    Checksum (SHA256): 64788bc774ae3d41b9b0b7bf335eb8cf0ba76c20dee69ebab763052b2ae08270
  Signatures: The parties sign by E-Signature (Swedish Bank ID), confirmation of digital signing will be added and
    tied to this agreement.
  Debtor signatory:
    Name: Nils Nilsson
    Id: 660101-1111
    Role: Signs in the role as board member
TechnicalDetails:
  ContentHash: "9b417e9a06c861570a22c83417fa27334c142801eb50055f5781952bdac0e515"
  OwnerKey: "03b3e0d980bb5454ffc83170812ad56d31a6b0b0d127dc7114c1d26f7f191db382"
  VersionKey: "a3491861825e01615a003660c0585f1c7601acadc104e05d75fd22afc44f46ed"
  Timestamp: "2019-05-20T15:21:57.728Z"
  TraceOriginalId: "23a01cbb62afa6a1a19f8effeb00c1cbdaddb07937ebf87570606351c0ad5399"
  LedgerSignature: "e9A/iGCUhqmsvploQTj9/fH882e5DVClNC4QcAn7fiV7FJ1X2gWxUNtmDs1p3lDyEZ2YZDPrlG491m3cR1DDJg=="
Amendments:
  -
    Content: |
      e-Signature of Nils Nilsson logged here
    TechnicalDetails:
      ContentHash: "62059ce894be081dba2f8ffe75d48c8166fa794e3d5ae8d1996c22aed88a5824"
      OwnerKey: "03b3e0d980bb5454ffc83170812ad56d31a6b0b0d127dc7114c1d26f7f191db382"
      VersionKey: "bfe4067749dee71914d811a132ad958c3c68625cd144c5b397f170de19877ed4"
      Timestamp: "2019-05-20T15:23:10.736Z"
      LedgerSignature: "LTPDldOyuv+KBi+q/OyZL/CFGp/SPL74UtfBpotgp5/L40QDGn310FQgox/TnlETTAgdtgjAwVZENWQA8zqvuw=="
```


Fig. 6.2 Promissory note realised as a Ricardian contract (Source: *trace:original* system (Enigio Time AB 2019))

created, containing the initial cryptographic envelope and the amendment content, i.e., the transfer information. The newly-created envelope containing information about the new owner is created and registered in the blockchain. The previous owner can retain his original but can no longer enjoy any associated rights because they have been transferred to the new owner (Fig. 6.3).

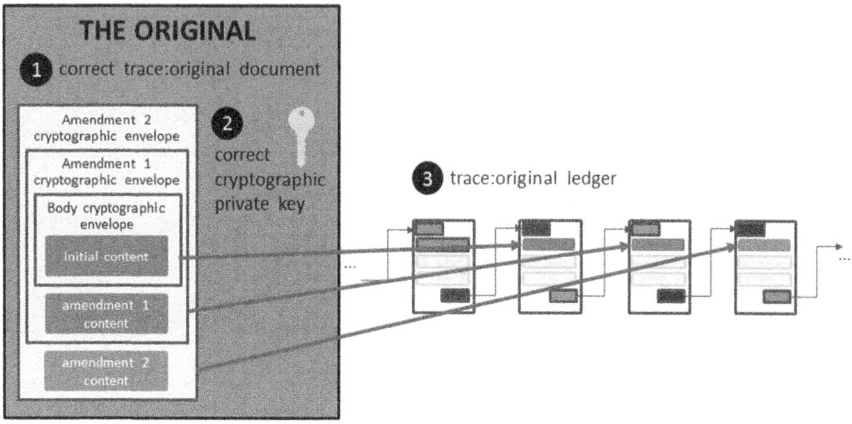

Fig. 6.3 Management of digital originals (Enigio Time AB internal documentation)

A simplified example of the ledger is shown in Table 6.1, with only the initial five letters of the hash values shown and several columns omitted (these contain additional signatures and other ledger information and are indicated by the "..." column header). The ledger is realized as an append-only structure. The following discussion will focus on the content indicated in bold.

The "Sequence ID 1" row shows that the document (e.g., a promissory note), having FD3F1 as the starting part of its calculated hash value, has been created (operation type "CRE"). It is in its first version and the owner's public key begins with AC21F. Anyone in possession of a copy in the form of an original can check against the ledger the integrity of the copy by calculating its hash and comparing it with the one in the ledger, but only the owner can manage the digital original. The next row shows the creation of another document, DF231.

The "Sequence ID 3" row indicates registration of an amendment (operation type "AME") to the first document (FD3F1), but the owner is still the same (AC21F). Note that the content hash for document FD3F1 changes because a new cryptographic envelope is created, encapsulating the previous one and the amendment information (e.g., a repayment has been collected). The next row shows the transfer (operation type "TRA") of the promissory note from one owner (AC21F) to another (FD3ED). From that moment, only the new owner (FD3ED) has control over the digital original and the rights coming out of that promissory note. Once the note's conditions have been met, the promissory note can be invalidated, but only by the current owner (FD3ED). The "Sequence ID 5" row shows invalidation (operation type "INV") of the digital original and the owner information is set to 00000. This means that there is no owner anymore, the promissory note has been invalidated, and no one has any rights coming out of it; however, all previous owners, just as anyone else who previously held a copy, can still be in possession of the promissory note as a record.

Table 6.1 Distributed ledger transactions

Sequence ID	Time stamp	Document ID	Content hash	...	Operation type	Document version	Public key	Ledger seal
1	2019-06-07 10:01:24	**FD3F1**	**FD3F1**	...	**CRE**	**1**	**AC21F**	14A2B
2	2019-06-07 10:01:26	DF231	AD121	...	CRE	1	EDF12	AE234
3	2019-06-08 13:23:12	**FD3F1**	DFF23	...	**AME**	**2**	**AC21F**	F4BC1
4	2019-06-09 15:09:45	**FD3F1**	A3E4E	...	**TRA**	**3**	**FD3ED**	5E4C1
5	2019-06-10 08:35:11	**FD3F1**	BA4DA	...	**INV**	**4**	**00000**	A2C4D

The analysis of this system for managing digital originals, using the example of negotiable instruments, shows that, although the digital medium enables creation of an original and its multiplication in an infinite number of identical copies, blockchain technology makes it possible to create a digital original and to identify it among other identical copies. Further, the example illustrates that the digital original can be in the sole control of the current owner who can prove their ownership and claim to the rights embedded in the original document. The append-only structure of the blockchain enables nonrepudiation, i.e., no party can deny signing of a contract or that a transaction has been made. Such a structure also allows for digital originals to be amended and ensures that originals cannot be changed without a trace. Therefore, the append-only structure creates and preserves an immutable audit trail. The owner can distribute as many copies as needed and everyone can verify against the ledger whether a copy corresponds to the current digital original. Also, the owner can transfer ownership of a digital original or invalidate it. However, it should be pointed out that the digital original itself can be stored wherever the owner prefers so that, if needed, all business details can be held as confidential.

To conclude, by taking a blockchain-based approach one can digitize paper processes, like the one with the negotiable instruments, where the key requirement is creation, management and control of digital originals and their differentiation from the identical digital instances. More importantly, the archival science requirement of preserving the trustworthiness of the digital original records is met—they can be assessed as accurate, reliable, and authentic.

6.5 Preservation of Provenance

Blockchain is a technology created to deal primarily with transactions. According to Duranti (1998), a transaction is "prompted by an act or will aimed to [...] create, maintain, modify or extinguish situations. Basically, when we refer to transactions and their products/objects, we are referring to records that should be compliant and preserved according to regulations and laws. Also, transactions are related to rights usually represented and claimed through those records and that is why it is so important to preserve such records through space and time. This section aims to present a brief overview of the design choices related to preservation of blockchain records that will also impact the durability of the systems based on blockchain platforms.

According to Lemieux (2016), preservation takes place over a time scale during which technologies, formats and preserving communities are very likely to change. In archival science and other information sciences the challenge of preserving digital resources is a subtle topic exhaustively discussed and yet with no perfect solution. One definite point of convergence is that there is no other way of preserving digital information than to start from the beginning, during the process of building the

Table 6.2 Metadata required to control different digital objects

What is in control	Suggested requirements
Transactions	ISO 20022, ISO 11179, e-GSM (UK)
Records/information	ISO 23081, MoReq 2010
Assets	ISO 19115, INDECS

solution for information keeping. In the case of records, we emphasize that the new complex and interactive forms of records require solid information on provenance to ensure their authenticity for future generations.

Preserving the provenance of digital objects increases trustworthiness. For that purpose, they must be managed in a way that secures and preserves knowledge of their origins and the circumstances of their creation (Yeo 2013), an idea strongly defended not only by records professionals but by different professions in business and technology fields. In the case of blockchain systems, as in the majority of systems,, one of the most common solutions to preserve digital objects and guarantee the sustainability of systems involves the use of metadata attached to the object of preservation. There are different specifications regarding different objects and context, as outlined in Table 6.2.

This is not an exhaustive list; however, the standards presented in the table could contribute to blockchain systems design and improve the quality and trustworthiness of the emerging technology. The point to be emphasized in this section is that no matter what we aim to preserve in the digital environment, the provenance of the digital object must feature prominently. *Metadata* is what makes it possible to identify all the events related to an object from its origin to its actual state, so thinking about metadata schema or data models when designing blockchain systems is an essential step in the long-term preservation of the information that such systems convey.

6.6 The Use Case in Health

Earlier sections of this chapter have discussed the use of blockchain as a potential *substitute* to centralized systems; explored how blockchain technology can be used to mitigate some issues in differentiating *authentic originals* from copies in the digital environment; and to *preserve provenance*, which is crucial for ensuring long-term access to, and determination of the authenticity of, information in the digital environment. In this section, we focus on the health sector as an emerging use case for blockchain technology.

Globally, there is a significant pressure on healthcare providers as the volume of user information (including data from IoT devices) and the variety of records continue to rise in the digital environment (Yasri 2018). Telemedicine, a method of health service delivery at distance, is using wearable devices, smart phones, and IoT devices in chronic health management, prescription compliance and collating real-time conditions. While these innovations are helpful in increasing interoperability, thus reducing administration inefficiencies, they also expose healthcare management to hacking, identity theft, and misuse of personal data. As Yasri (2018) notes, the data stored by health providers are subject to breaches, compromising the security and integrity of the information. Besides that, manipulation of data causes insurance frauds, duplicate claims, and billing for services that were not rendered. Blockchain can help to ensure secure transmission of data as well as ensuring that the data are safely stored. As Unnithan et al. (2020) ascertain:

> [the] issue of data security could be addressed when the encrypted distributed ledger blockchain technology can be used for the safe and immutable transmission of electronic health record data. While issues of hacking and ransomware attacks are prevalent within the health sector, a suitably implemented blockchain technology could potentially enable patient health data to be shared while preserving data security and integrity.

Unnithan et al. (2020) explain that the feature that differentiates blockchain security from existing health data sharing approaches (such as Health Information Exchanges) is the ability for all parties in the network to validate the stored data, which is fully automated. On the one hand, there is built-in hashing encryption used in blockchains for secure transmission of data. Security is further heightened by the fact that every node on the network will have a copy of the blockchain. Effectively, if hacking or fraud is attempted, the node sending the block that contains the fraudulent data will be rejected by the wider network (Unnithan et al. 2020).

How can data provenance be traced with blockchain? It is now proven that blockchain software can help track the origin of, and subsequent changes to, patient data including medical records, imaging data, test results etc. (Unnithan et al. 2020). The software may deny or approve access to data only to authorized parties in the blockchain in specific geographic locations, allaying concerns about jurisdictional issues of data transfer. When technical failures such as loss of power are considered, it may be noted that all nodes in a blockchain network cannot collapse at the same time, meaning that the network will continue to function on other nodes. And if data is securely and suitably accessed, any pertinent changes to evidence would be visible immediately. These aspects of the technology ensure that there is an ability to trace provenance.

In the previous sections, we introduced the concept of smart contracts and Ricardian contracts. In the health context, smart contracts can facilitate storage of health records and information of patients (users). When users move locations, they may still allow their preferred healthcare providers (GPs) to view their records utilizing smart contracts running over a blockchain network. Enhanced interoperability and reconciliation are facilitated in such environments, i.e., recognising any user as themselves, across any geographic location. Conversely, when a network is notified of a consultation or any service, insurance payments can be released to relevant entities. Smart contracts assist in compliance and adherence to standards across geographic jurisdictions with real-time updates.

The World Health Organization (WHO) (2016, pp. 21–28) recommends that every country should work towards implementing Electronic Health Records, to provide citizens with improved healthcare services and delivery efficiencies. One of the main impediments to this implementation is the lack of trust between healthcare institutions (Hripcsak et al. 2014). Sharing of data sets across geographic jurisdictions may be limited due to lack of interoperability standards making integration and interpretation of data from disparate systems difficult (Unnithan et al. 2020). A decentralized blockchain system would mean that all user medical data could be connected, with patients' informed and validated consent, and referenced by healthcare professionals. Effectively, the application of blockchain technology has the potential to ensure data are not only secure but also, in real time, consistent across numerous digital platforms, thereby enhancing *social trust*.

Thus, in the sensitive environment of health, security of transmission and tracing of provenance are both possible via appropriately facilitated blockchain technology. When users have trust in the system, i.e., they have control over their data and its secure transmission, the concept of *social trust* is enhanced, enabling a national-level Electronic Health Record facilitation, as recommended by WHO.

Various pilots involving the application of blockchain to healthcare research and service delivery have occurred in the 2017–2019 period. Estonia, for example, has implemented blockchain, led by the national government, for secure protection of Electronic Health Records (e-Estonia 2018, 2020). It may be noted that the country's health information was integrated through a centralized system, which allows for seamless access and data sharing across all providers, while giving users the ability to control their health data (Novek 2018).

In Australia, the federal department of health ran a pilot to use blockchain for medical research records, and provide an immutable record for tracking health data search queries and downloads—allowing the researcher to protect his/her intellectual property associated with their proven hypotheses, whilst protecting highly sensitive data in alignment with the government's Data Access and Release Policy (Australian Government 2018). The technology enables governments to make de-identified and confidential patient health records available to the research community in a way that fosters *traceability* and reproducibility of research results. As such, researchers need to justify and ensure the reproducibility of their search queries and results along with their research method. Essentially this means ensuring data integrity (i.e., the search query and retrieved data are not modified) and non-repudiation, i.e., proof that the data is time stamped based on a specific query. Blockchain has proven to be successful in this venture.

A blockchain-based notarization pilot implementation to address data integrity and non-repudiation in biomedical research was presented by Kleinaki et al. (2018). The 'Smart Digital Contract' tool has been tested on the Ethereum blockchain platform and is able to effectively query and retrieve data using a third-party notary service on two major biomedical databases PubMed MEDLINE and CARRE risk factor reference repositories. They recommend this method for ensuring data integrity and non-repudiation (e.g., using digital signatures) which can be combined with blockchain features to enhance contract traceability.

6.7 The Use Case in Banking and Finance

The concept of a digital original has been realized by the Stockholm-based blockchain innovation company Enigio Time AB as a service called *trace:original* (Enigio Time AB 2019). Their focus is on the mortgage and corporate lending and trade finance areas as well as, to a lesser extent, logistics. For mortgages and corporate lending, their solution most commonly addresses promissory notes. For banks, the motivation to go digital in this area is partly cost reduction but mostly to enhance the customer experience and provide the possibility to do remote signing for the mobile generation. Enigio estimates that, on average for mortgages in Europe, the cost of a physical mortgage during its lifecycle is somewhere between 400 and 800 €. By applying the concept of a digital original through the *trace:original* solution, they claim the cost can be reduced by at least 80% and, most importantly, can give customers a better experience.

In the area of mortgage and corporate lending Enigio is currently implementing the *trace:original* solution with Stabelo (a mortgage institute), and DBT Capital (a capital lending provider). Having digitized the process with promissory notes those two companies can have a complete digital process where no paper documents or handwritten signatures are needed. Customers do not need to come to Stockholm to sign agreements or send signed papers back and forth. In addition, since the asset is in the digital form, it is possible to verify and reconcile the data computationally at any given moment. This is also important for the secondary market—a batch of digital originals can be sent over before the transaction is made and the ownership and the data can be verified instantly.

Within trade finance Enigio have, in cooperation with the International Trade and Forfaiting Association (ITFA), launched a pilot project where the *trace:original* solution will be tested for digitization of documents such as bills of exchange, promissory notes, and guarantees. The pilot participants include Lloyds Bank, SMBC, Crown Agent Bank, China Systems, Finastra, and many others. In addition to paving the way for a complete digital trade finance process, this could also prevent fraud which is, in general, a significant problem within trade finance. In the area of logistics, the *trace:original* solution can be used for any type of document where the possession of an original is essential, e.g., shipping documents, inspection certificates, etc.

6.8 Conclusion

In this chapter, we have illustrated how using blockchains to establish accurate, reliable, and authentic provenance records can transform social trust in two key application areas: business and healthcare. The relationship between technical features of blockchain and social trust is not a direct one; rather, it passes through and relies upon accurate, reliable, authentic and preserved data and records because social trust depends upon possessing trustworthy information about the thing or person in which trust is to be placed. Thus, though it may seem unrealistic for a technical system, especially an emerging one, to effect a transformation in trust among social actors, in the examples we have discussed in this chapter we point to the unique capabilities of blockchain systems—to incentivize and enhance accuracy and reliability of records, protect their authenticity, and trace the provenance of data and records—that have the potential to create the necessary informational foundation for the transformation of social trust.

References

Australian Government, Department of Health. (2018). *Data access and release policy*. Retrieved from http://www.health.gov.au/internet/main/publishing.nsf/Content/Data-Access-Release-Policy

Business Dictionary. (2019). *Negotiable instrument*. Retrieved August 25, 2019, from http://www.businessdictionary.com/definition/negotiable-instrument.html

Duranti, L. (1998). *Diplomatics: New uses for an old science*. Lanham, MD: Scarecrow Press.

e-Estonia. (2018). *Healthcare*. Retrieved from https://e-estonia.com/solutions/healthcare/e-health-record/

e-Estonia. (2020). *KSI® blockchain in Estonia*. Retrieved from https://e-estonia.com/wp-content/uploads/2020mar-faq-ksi-blockchain-1-1.pdf

Enigio Time AB. (2019). trace: original. Retrieved August 25, 2019, from https://www.enigio.com/traceoriginal

Grigg, I. (2004). *The Ricardian contract*. In B. Benatallah, C. Godart, & M.-C. Shan (Eds.), *WEC 2004: First IEEE International Workshop on Electronic Contracting,* (pp. 25–31). Los Alamitos, CA: IEEE Computer Society. https://doi.ieeecomputersociety.org/10.1109/WEC.2004.1319505

Gupta, A. (2009). Data provenance. In L. Liu & M. T. Özsu (Eds.), *Encyclopedia of database systems*. Boston, MA: Springer. https://doi.org/10.1007/978-0-387-39940-9_1305.

Hofman, D., Batista, D., & Lemieux, V. L. (2018). *Centre of excellence for prevention of organ failure (PROOF) – (RPCCA-01) – Case Study 1*. University of British Columbia Records in the Chain Project. Retrieved August 3, 2019, from http://blogs.ubc.ca/recordsinthechain/files/2018/06/PROOF-Case-Study_22-June_FINAL.pdf

Hripcsak, G., Bloomrosen, M., Flatelybrennan, P., Chute, C. G., Cimino, J., Detmer, D. E., Edmunds, M., Embi, P. J., Goldstein, M. M., Hammond, W. E., Keenan, G. M., Labkoff, S., Murphy, S., Safran, C., Speedie, S., & Wilcox, A. B. (2014). Health data use, stewardship, and governance: Ongoing gaps and challenges: A report from AMIA's 2012 health policy meeting. *Journal of the American Medical Informatics Association, 21*(2), 204–211. https://doi.org/10.1136/amiajnl-2013-002117.

InterPARES Project. (2001). Authenticity task force report. In L. Duranti (Ed.), *The long-term preservation of authentic electronic records: findings of the InterPARES project*. Vancouver, BC: InterPARES Project. Retrieved August 25, 2019, from http://www.interpares.org/book/interpares_book_d_part1.pdf

Kifor, T., Varga, L. Z., Vazquez-Salceda, J., Alvarez, S., Willmott, S., Miles, S., & Moreau, L. (2006). Provenance in agent-mediated healthcare systems. *IEEE Intell Syst, 21*(6), 38–46. https://doi.org/10.1109/MIS.2006.119.

Kim, H. M., & Laskowski, M. (2018). Toward an ontology-driven blockchain design for supply-chain provenance. *Intelligent Systems in Accounting, Finance and Management, 25*(1), 18–27. https://doi.org/10.1002/isaf.1424.

Kleinaki, A.-S., Mytis-Gkometh, P., Drosatos, G., Efraimidis, P. S., & Kaldoudi, E. (2018). A blockchain-based notarization service for biomedical knowledge retrieval. *Computational and Structural Biotechnology Journal, 16*, 288–297. https://doi.org/10.1016/j.csbj.2018.08.002.
Lemieux, V. L. (2016). Provenance: Past, present and future in interdisciplinary and multidisciplinary perspective. In V. L. Lemieux (Ed.), *Building trust in information* (pp. 3–45). Cham: Springer International.
Lemieux, V. L. (2017). *Blockchain and distributed ledgers as trusted recordkeeping systems.* Presented at Future Technologies Conference 2017, Vancouver, BC, November 29. Retrieved from https://saiconference.com/Downloads/FTC2017/Proceedings/4_Paper_279-Blockchain_and_Distributed_Ledgers_as_Trusted.pdf
Nayak, K., Kumar, S., Miller, A., & Shi, E. (2016). Stubborn mining: Generalizing selfish mining and combining with an eclipse attack. In *Proceedings of the 1st IEEE European Symposium on Security and Privacy* (Euro S&P 2016) (pp. 305–320). Los Alamitos, CA: IEEE Computer Society. Retrieved from https://ieeexplore.ieee.org/document/7467362
Novek, A. (2018). *Blockchain in Estonian National Health Information System* [PowerPoint slides]. Retrieved from https://cdn.ymaws.com/echalliance.com/resource/resmgr/images/DHW_2018_Profiles_/2018_Presentations_/Artur_Novek.pdf
Pearce-Moses, R. (2005). *A glossary of archival and records terminology.* Chicago: Society of American Archivists. Retrieved August 25, 2019, from https://www2.archivists.org/glossary
Provenance. (n.d.) *Multilingual archival terminology database.* Retrieved August 13, 2019, from http://www.ciscra.org/mat/mat/term/283
Provenance, n. (n.d.). *OED Online.* Retrieved from https://www.oed.com
Unnithan, C., Houghton, A., Anema, A., & Lemieux, V. (2020). Blockchain in global health – An appraisal of current and future applications. In L. Kuan-Ching, C. Xiaofeng, J. Hai, & E. Bertino (Eds.), *Essentials of blockchain technology.* Boca Raton. FL:CRC
Wand, Y., & Wang, R. Y. (1996). Anchoring data quality dimensions in ontological foundations. *Communications of the ACM, 39*(11), 86–95. https://doi.org/10.1145/240455.240479.
Wang, R. Y., Storey, V. C., & Firth, C. P. (1995). A framework for analysis of data quality research. *IEEE Transactions on Knowledge and Data Engineering, 7*(4), 623–640. https://doi.org/10.1109/69.404034.
World Health Organization (WHO). (2016). *From innovation to implementation: eHealth in the WHO European region.* Retrieved from http://www.euro.who.int/__data/assets/pdf_file/0012/302331/From-Innovation-to-Implementation-eHealth-Report-EU.pdf
Xu, S., Rogers, T., Fairweather, E., Glenn, A., Curran, J., & Curcin, V. (2018). Application of data provenance in healthcare analytics software: Information visualisation of user activities. *AMIA Joint Summits on Translational Science Proceedings, 2018*, 263–272. https://www.ncbi.nlm.nih.gov/pmc/articles/PMC5961786/
Yasri, D. (2018). Bringing blockchain technology to telemedicine. *Medium.* Retrieved from https://medium.com/pikciochain/bringing-blockchain-technology-to-telemedicine-4090d283922b
Yeo, G. (2013). Trust and context in cyberspace. *Archives and Records, 34*(2), 214–234. https://doi.org/10.1080/23257962.2013.825207.

Chapter 7
Conclusion: Theorizing from Multidisciplinary Perspectives on the Design of Blockchain and Distributed Ledger Systems (Part 2)

Victoria L. Lemieux and Chen Feng

7.1 Introduction

Models play an important role in the sciences as devices for scientific discovery (Bailer-Jones 2003; Eckert and Hillerbrand 2018). In the introduction to this volume we discussed how our multidisciplinary strategic design workshop began with consideration of a "three layer" model of blockchain and distributed ledger technology (DLT), which we used as a framework to explore five key themes commonly associated with such systems: governance, incentives, security, decentralization, and provenance.

At the outset of our collective intellectual journey, the interactions among these layers[1] were not sharply defined. As our strategic design process and collective chapter writing progressed, however, the scope of each layer and the interactions among them came into sharper focus. We begin this chapter with a discussion of

[1]There was some discussion in the group as to whether these layers should be called dimensions rather than layers. The argument for the use of the term dimensions centered on a concern that it would be confusing for software developers who typically conceptualized of layers as layers in the TCP/IP stack. On the other hand, some saw the use of the term, and its association with software development, as being advantageous in promoting an extended understanding of the software development stack as expanding into social and data/records design considerations. The group came to no definitive conclusion on this issue of nomenclature but, as discussed in this chapter, a deeper understanding of the layers as subsystems of DLT systems has emerged from a synthesis of multidisciplinary perspectives.

V. L. Lemieux (✉)
School of Information, University of British Columbia, Vancouver, BC, Canada
e-mail: v.lemieux@ubc.ca

C. Feng
School of Engineering, University of British Columbia (Okanagan Campus), Kelowna, BC, Canada
e-mail: chen.feng@ubc.ca

V. L. Lemieux, C. Feng (eds.), *Building Decentralized Trust*,
https://doi.org/10.1007/978-3-030-54414-0_7

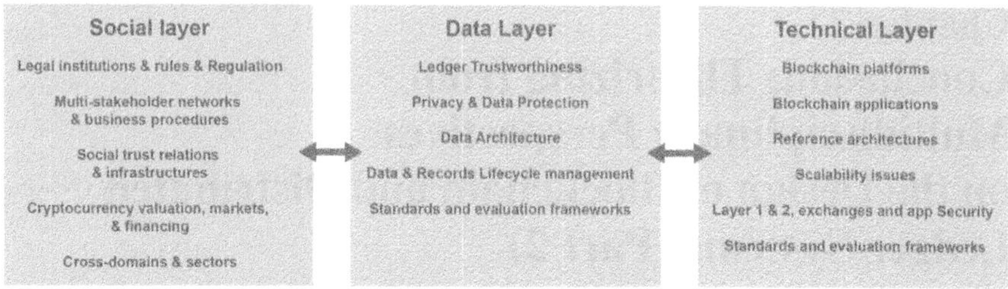

Fig. 7.1 Three-layer trust model of DLT (Lemieux et al. 2019)

ontological foundations. In the second section of this chapter, we draw upon systems theory to present refinements to the model emerging from our multidisciplinary discussions. In presenting these refinements, our goal is not only to offer a descriptive model of DLT systems, including blockchains, but also to offer a model to aid design of such systems, which we address in the third section of this chapter. The overall goal of this chapter, then, is to articulate a scientific model of DLT systems that has the power to describe a range of such systems as well as an engineering design model that has evaluative power to assess multiple design alternatives against a set of requirements and generative power to help designers create new (and better) designs that satisfy identified requirements (Beaudouin-Lafon 2004; Meyer et al. 2015; Staples 2014).

7.2 Ontological Foundations

In Chap. 1, we presented a visual representation of the simple "three layer" model (see Fig. 7.1) with which we began our collective multidisciplinary exploration. Every model expresses one or more theories of some sort and, "in many scientific contexts, models are central epistemic tools that may not be subordinate to theories..." (Eckert and Hillerbrand 2018, p. 220). Our model is no different in this sense in its expression of a theory of DLT systems as *socio-informational-technical systems*. Taking this ontological stance, social and information aspects of DLT systems do not just use or inform system design, as in traditional engineering conceptualizations, but are part of the operation of the system, i.e., they are sub-systems within the larger system, not an assemblage of merely technical components. This is closer to a "mixed initiative" or "human-in-the-loop" conceptualization of systems, in which human actors and computational components interact to create an intelligent machine (Haller et al. 2013; Harris 2018).

In thinking of DLT systems this way, we draw upon the work of Latour (1986, 1987, 2005) who argues that any technology comprises an "assembly of forces" and a system of alliances (see also Bousquet 2014). In this system of alliances, Latour

(1986, 2005) placed no greater importance in physical objects than social actors, or "actants". Each he considered to be ontologically equivalent (Orlikowski 2007). We adopt this ontological position without implying that socially constructed reality is completely unhinged or disconnected from physical reality, or that social components and physical components possess isomorphic characteristics. It is a world in which a "physical" ontology, such as Bunge's (2012), operates in parallel to a social ontology, such as Searle's (2006), and an information ontology, such as Floridi's (2005). We argue that this is a stance likely to generate greater understanding when exploring and assessing DLT systems, given that such systems touch on issues of social trust, documentary representations of socio-institutional power in action (aka records), and computer information processing. Latour asks us to examine what alliances—whether social, informational, or physical—have to be formed in the operation of a machine and what interests had to be negotiated (Latour 2005; Orlikowski 2007). Our "three layer" model is intended to prompt the same questions.

Given this ontological stance, we reject pure materialism which suggests that only physical objects are "real" and can have effects in the physical world. A border crossing manned by border guards, which is a geo-political creation, under the right conditions, has as much power to stop a person from crossing from one geo-physical space to another as a river or a wall and, in addition, the physical reality of the existence of a human social actor may change quite dramatically upon crossing from one constructed socio-political reality into another. Similarly, we reject pure social constructionism, since we admit that engineering artefacts, such as DLT system designs and actual DLT systems, are as capable of constructing social reality as are social actors themselves. Indeed, for engineers, that is the point of creating such systems.

We accept as true that systems embed the "will" of social actors, who we recognize as having agency, and thus, act as indirect mechanisms of social construction, but we also note that this is not a perfect, noise-free zone of will transmission. In other words, the requirements of system users, from whom designers may collect requirements for DLT systems, may not perfectly transmit their requirements to designers and designer's conceptualizations of DLT systems may not perfectly reflect users' requirements in their designs. Moreover, we also recognize that an evolving system may create its own future possibilities, or what Kauffman refers to as "unprestatable opportunities that emerge in an unprestatable ever-growing and changing adjacent possible that [social actors] partially co-create, with and without intent." (2013, p. 22). We accept and embrace the productive tension that emerges from the aim of modelling and theorizing about DLT systems and the potentially unknowable reality of DLTs as complex socio-informational-technical systems.

This is most evident in the fact that, even if designs are perfect representations of users' requirements and designers perfectly represent users' requirements, users of systems may not use the designed artefact (i.e., DLT systems) as intended.

An ontological stance that places physical, informational, and social constructs on equal ontological footing still allows us to retain a critical rationalist approach to engineering design which builds upon Popper's three worlds model (Popper and Eccles 1977). In the three worlds model, as Staples (2014) explains:

> World 1 is the world of physical entities and phenomena. World 2 is the world of mental or subjective states and events. World 3 is the world of objective content: knowledge and products of thought that can be explicitly recorded or spoken. The worlds are not distinct, because mental states have a physical basis and because objective knowledge can be understood and can be physically represented. There are direct interactions between World 1 and World 2 (e.g. sense perception and the will to act); and between World 2 and World 3 (e.g. representation and understanding). Interactions between World 1 and World 3 (e.-g. prediction of empirical phenomena by theory) are only indirect. They are mediated by the second World of human understanding and intention (p. 13).

We can consider a DLT system as a design artefact (World 3) that represents a real-world use case, mediated, for example, through the mental and subjective states of the designer and social actors (World 2), documentary representations of social constructs, such as policies and procedures (World 3), and social actors' lived experiences (what Popper sometimes treated as a separate world). This stance also allows for formalizing of engineering design as Staples (2015) outlines:

> ...that for any state of the world x, including an artefact a, where acceptable environmental conditions E apply in the world and to the artefact, and where the artefact fits a design D, then requirements R will be satisfied. When applied to reason about a specific artefact, the requirements for the artefact must be within the theory's predicted performance of the artefact R, and the actually-acceptable limitations on the specific operating environment must contain the theory's environmental conditions E. Designs are usually abstractions, often expressed in a form that is consistent with relevant analytical theories. Multiple artefacts may satisfy a single design, and a single artefact may satisfy many designs. Formula 1 can be decomposed using *modus ponens*:
>
> $$[E(x,a);D(a)]' \ B(x,a) \tag{2}$$
>
> $$[E(x,a);D(a)]' \ B(x,a) \rightarrow R(x,a) \tag{3}$$
>
> Engineers may use one set of rules (formula 2) to predict artefacts' behavior B, then separately reason (formula 3) about how that behavior satisfies requirements R, thus deriving the overall claim (formula 1). This problem decomposition allows the development and use of generic theories to predict performance. The claims of a very general theory are unlikely to be identical to particular requirements specifications R, but may entail them. (p. 18)

The "critical rationalist" view of the world expressed in the above quote typically stands in opposition to "postmodernist" or sociological epistemological philosophies, but in relying upon Latour's ideas we believe we have found a possible way to integrate the ontological and epistemological collisions that inevitably occur in any multidisciplinary work in a way that advances the development of our "three layer" model, if not yet in a way that is completely logically coherent. Logical coherence remains future work.

7.3 Refining the "Three-Layer" Model as Description of DLT Systems

Owing to its power as a meta-disciplinary theoretical framework (Checkland 1999), our model adopts general systems theory (Von Bertalanffy 1950, 1968) as an underlying theoretical framework in conceptualizing of DLTs, including blockchains. Not only may DLTs be characterized as systems, they may also be characterized as *complex systems*; that is, "systems that do not have a centralizing authority and are not designed from a known specification, but instead involve disparate stakeholders creating systems that are functional for other purposes and are only brought together in the complex system because individual agents of the system see such cooperation as being beneficial for them" (Sheard and Mostashari 2009, p. 296).

A system is comprised of interacting sub-components that are interconnected through a web of relationships. Each component—in the case of DLTs, the social, data/records and technical sub-components—functions independently and collectively as a part of the system toward achieving a single purpose (Midgley 2003; Tejeida-Padilla et al. 2010), and they are dynamically interrelated and interdependent (Skyttner 1996, p. 30) as we have discussed in the preceding chapters. Effective operation of the parts in relation to the whole leads to the achievement of the system's purpose (Ackoff 1994; Midgley 2003; Senge 2006). In the case of DLTs, we conceptualize of them as systems that function with a purpose of achieving trust among social actors (Vigna and Casey 2019). Trust we define as the degree to which a user or other stakeholder has confidence that something—e.g., a person, product, or system—will behave as intended (International Organization for Standardization [ISO], 2020). Yet, as several chapters in this volume discuss, even with this definition, trust is a complex, multi-dimensional concept, making it challenging to use in system design and evaluation. To illustrate the multi-dimensionality of the concept, trust in the context of dependable software systems usually refers to a willingness to accept a dependence upon someone or thing (Rousseau et al. 1998), rather than solely having confidence in that person or thing. Yang et al. (2016) identified the following three categories of trust for the credibility of a peer-to-peer (P2P) lending platform: (1) system-based trust through service quality (efficient and flexible transactions); (2) cognitive-based trust, such as first impressions through awareness, reputation and addressing perceived risk; and (3) affective-based trust, such as the utilization of social networking supporting long-term strategic alliances. Despite the challenging complexity of the concept of trust, there is nevertheless broad agreement that trust among social actors is essential for the effective and efficient functioning of social systems. Without trust, we cannot have confidence to interact with one another for social, economic, or political purposes. Social interactions become expensive without trust, introducing high institutional transaction costs (Coase 1937; Williamson 1979, 1986) at best, or, at worst, cause social interactions to simply grind to a halt.

Some may find our argument that DLT systems operate to achieve a purpose of social trust as too narrow a conceptualization. After all, if DLTs are used to establish the identities of interacting autonomous vehicles, trust would be established between the interacting vehicles, and only indirectly between social actors. We maintain, however, that though trust in this case is between interacting non-human entities, it ultimately resolves to, and results in, greater social trust. In this case, the use of DLT to identify distributed autonomous vehicles affords a higher level of system security (e.g., it serves to prevent untrusted entities from interfering with the operation of the network). This higher level of technical security gives social actors relying upon the network of autonomous vehicles greater confidence when relying upon using it. This, in turn, generates the opportunities afforded by networks of autonomous vehicles for greater efficiencies in social interactions (i.e., reduced transaction costs). Similarly, DLT systems designed to improve the trustworthiness of recordkeeping systems, such as land registers, or that aim to create records that trace the provenance of physical assets such as diamonds, artworks or food, achieve the aim of social trust indirectly by means of creating systems of recordkeeping that provide social actors with the knowledge to have confidence in relying upon others to act (i.e., to trust that the individual from whom they wish to purchase a piece of land actually holds title to that land, or that the Red Snapper they wish to purchase is really that kind of fish and not another.) Indeed, in DLT systems, a distributed ledger is a key system output, and thus it can be argued that social trust is always mediated through an informational (data/records layer). In turn, distributed ledgers are instantiated through computational technologies such as algorithms, networks, servers, etc., and thus social trust is arguably always mediated in some way through the technical (in the sense of information and communication technologies) as technical components operate to instantiate the informational (i.e., data and records) layer. Thus, though a purpose of a DLT system is social trust, social trust may be achieved only through attending to trust at the technical and data/records layers as well, which is, again, why we argue that DLTs are socio-informational-technical systems.

Though we accept that there are three layers and that they all interact and dynamically shape one another, there remains an open question about the ordering of the layers. In our model, it may seem as though we have given some pre-eminence to the social layer by asserting that a goal of DLT systems is social trust. However, in our final version of the model, we separate the system goal of social trust from the social sub-system/layer of the DLT system. The social sub-system operates together with the other sub-systems/layers to achieve social trust. Indeed, as DLT systems instantiate an immutable ledger, there is a case to be made for according the data/records layer considerable importance as well. And, undeniably, DLT systems are inherently technical, in the sense of relying upon computer technology. Thus, we do not present any layer as any more or less important than any other layer. In Chap. 3, the authors show a model that has data passing through a technical layer to social actors. In this model, *technology* is the mediating layer, not information or data/records. Much more work is needed to understand which layers mediate the others, when, and by what means, and whether this is a stable or dynamic relationship.

Hence, in our final version of the model, we favour presenting the layers simply as interacting sub-systems of the DLT system as a whole, rather than ordered layers.

Consistent with the notion of a system as a set of interrelated components that function together within constraints towards a common purpose (Bittel 1978), social trust is not only a purpose of a DLT system, it operates as a system constraint (and thus as the ultimate requirement). In ecological interface design or EID (e.g., Burns and Hajdukiewicz 2004), a system constraint is necessary to identify a system's "Space of Permissible Actions", i.e., what the system is and should be permitted to do. The EID approach, though developed for interface design in the context of Cognitive Systems Engineering or CSE (Rasmussen et al. 1994; Vicente and Rasmussen 1990; Woods and Roth 1988), which addresses the issue of analyzing and designing process control systems from a human factors perspective, is well-suited to consideration of DLTs given that human and information aspects of system design figure prominently in the CSE approach as they must do in designing DLTs as socio-informational-technical systems. The objective of the EID approach is to enable the human operator to engage in adaptive behaviour if unanticipated or unexpected circumstances occur, thereby improving the overall safety of a human-machine system (Vicente and Rasmussen 1990, 1992). One way it seeks to achieve this aim is to address the Law of Requisite Variety (Ashby 1991), which essentially states that a system should be designed to support the variety of situations it is likely to encounter. In order to do this, it is therefore necessary to identify the Space of Permissible Actions, i.e., what the system is and should be permitted to do, and to make system boundaries and constraints directly visible in the user interface. If these boundaries and constraints are visible, as well as the current state of the system's performance, it becomes possible to control the system to more intelligently direct its operations. This task is complicated by the fact that all systems may generate emergent properties. It is possible for an infinite number of non-prestatable functions to exist for any given system (Kauffman 2013). These can give rise to a "web" of both causes and effects, as well as co-evolution of the system. The explicit identification of system goals/constraints and boundaries/space of permissible actions is helpful in abstracting away from the unknowable universe of functions and possibilities to a manageable range of functions and possibilities in developing a system design, and to design system behaviours that signal when a system enters a state outside of designed constraints and spaces of permissible actions.

Systems have control mechanisms. These may be internal, or endogenous, to the system, such as in the case of a temperature gauge on a heating system that senses the ambient temperature to adjust the system operation to match the desired output. Internal control mechanisms in the context of DLT systems are often portrayed as technical in nature, e.g., DLT consensus mechanisms that determine the rules for transaction or block confirmation for a given DLT system, which gives rise to the conceptualization of such systems as operating according to the "rule of code" as opposed to the usual socio-political configuration of governance (see, Chap. 2 for more on this theme). While internal control mechanisms, or sub-systems, in DLT systems do, indeed, operate by technical means, in our refined model, we portray internal control as a "wrapper" around the entire DLT system, and its social, data/

records, and technical sub-components. Rather than conceptualizing of governance as rule of code (technical) or rule of law (social), we suggest that rule of code as a form of internal governance or control interacts with all three sub-systems (i.e., technical, data/records, and social). We further conceptualize of it not as operating wholly by technical means, but rather by means of the design of interactions among the social, data/records and technical sub-systems that order the operation of the DLT system in a manner intended to drive the system toward homeostasis.

Control mechanisms may also be external, or exogenous to the system, yet still regulate it, as in the case of laws, regulations and standards, such as the Sarbanes-Oxley Act and the Control Objectives for Information and Related Technologies (COBIT) framework that stipulate rules for testing the operation of technology for financial accounting to ensure reliable accounting information (Lainhart 2000). In the case of DLT systems, external forms of governance typically take a social and human form, and thus are subject to human socio-political forces, i.e., the "rule of law" or "small p" politics.

Control mechanisms are typically conceptualized as "governance", with internal control being internal governance of the system—a form of self-regulation—and external control being those processes and regulations that provide for external governance of the system. Systems often integrate and rely on both forms of governance. For example, when internal governance mechanisms fail, there may be a need to revert to external governance mechanisms as a fail-safe mechanism and means of repairing or adjusting internal governance systems. A perfect example of this in the context of DLTs is the response to the DAO exploit of 2016, wherein human social actors had to step in to address a problem that arose from a technical internal control failure (DuPont 2019; Walch 2019a). This view allows for a theory of governance of DLT systems that transcends particular contexts; however, there is still need—as the authors of Chap. 2 argue—for a grounded approach to generating a theory of DLT governance. Research which undertakes this grounded theoretical work and that analyzes the results in relation to the efficacy of a general systems theoretic model of DLT system governance would add to the theoretical discourse presented in this chapter.

The boundaries of a DLT system comprise the structural and functional information about a DLT use case (i.e., a situation in which greater social trust is sought, either directly or indirectly) in a way that shows the means (interaction of the three layers) to ends (social trust) relationships that are embodied in it. The components of the system—our three layers—can be considered to be operating effectively when the goal, or requirement, of social trust is achieved.

However, we must also consider that people can use designed objects for alternative purposes; hence, design intentions may not necessarily correspond with actual designs (Kroes 2002; Staples 2015). A good example of this phenomenon is offered by Cornelius (2020) writing about the use of blockchains in supply chain management:

> Mac McGary of Sweetbridge, a nonprofit, open-source project that strives to "provide a set of rules, messages, and agreements that govern interactions of processes among humans, apps and machines," noted how smart contracts could be used to track and monitor the

> actions of rice farmers, possibly giving them control over their labor efforts and, at the same time, increase the production of the farms. While this project seems to have good intentions, if smart contracts are tracking and automatically manipulating the terms of a labor contract, and the actual atomized unit tracked is a human being instead of a currency, the literacy, comprehension, and awareness issues typically associated with previous standardized contracts need immediate attention. Additionally, the types of oversight needed to prevent abuse in this high-stakes and sensitive situation should be sorted out so as not to allow these types of contracts to exist behind the scenes, further exacerbating "black-box" culture. Similar to how the early proponents of cyberspace who imagined a world free of material consequences were disappointed by the utilization of this freedom by fascists, Nazis, and other types of trolls . . . the cypher-punks who idealize blockchain technology and cryptocurrencies might be similarly disappointed to see its offerings (i.e., anonymity, automation, trustless transactions) used in unintended ways. (p. 644)

Consequently, any measure of the efficacy of a DLT system would have to assess the degree to which the designer's intention had been realized in the actual design of the DLT system as well as the degree to which the DLT system achieves its purpose of social trust in operation over time. How to conduct these measures remains an open research question, however. At the very least, this work requires much deeper multidisciplinary exploration of the relationship between social trust as a goal and the design, implementation and use of DLT systems. Nevertheless, in the case of DLT systems, when social trust is achieved, it can be conceptualized as a state of system equilibrium or homeostasis.

Our refined model indicates that there is considerable dynamism and interaction among the three layers, or what we now would prefer to characterize as sub-systems, and that, in fact, most DLT systems do not achieve equilibrium, but are unstable. In fact, we may even question whether such systems are inherently unstable, given the dynamic nature of social trust (Fig. 7.2).

7.4 The Three Layer Model as a Framework for Design Alternatives

The thematic areas—governance, incentives, security, decentralization, and provenance—that we collectively arrived at as key DLT system characteristics can be viewed as a non-exhaustive set of elements that might be adjusted (traded-off) across the three sub-systems/layers to achieve DLT system equilibrium (i.e., social trust). Though each of these elements can in theory be given equal weight as an element of design, they are not equivalent in nature. Governance, for example, is a system control sub-system, while incentives are actuators of system control logic that form part of governance sub-systems. Security, on the other hand, signifies a broad concept covering a wide range of related concepts and DLT system properties, viz., data integrity, authenticity, accountability, non-repudiation, transparency, and confidentiality (which, in turn, may cover both secrecy and privacy). Decentralization is a state representing the architecture of a DLT system, or how the system is organized. Provenance is a capability made possible through designs that leverage

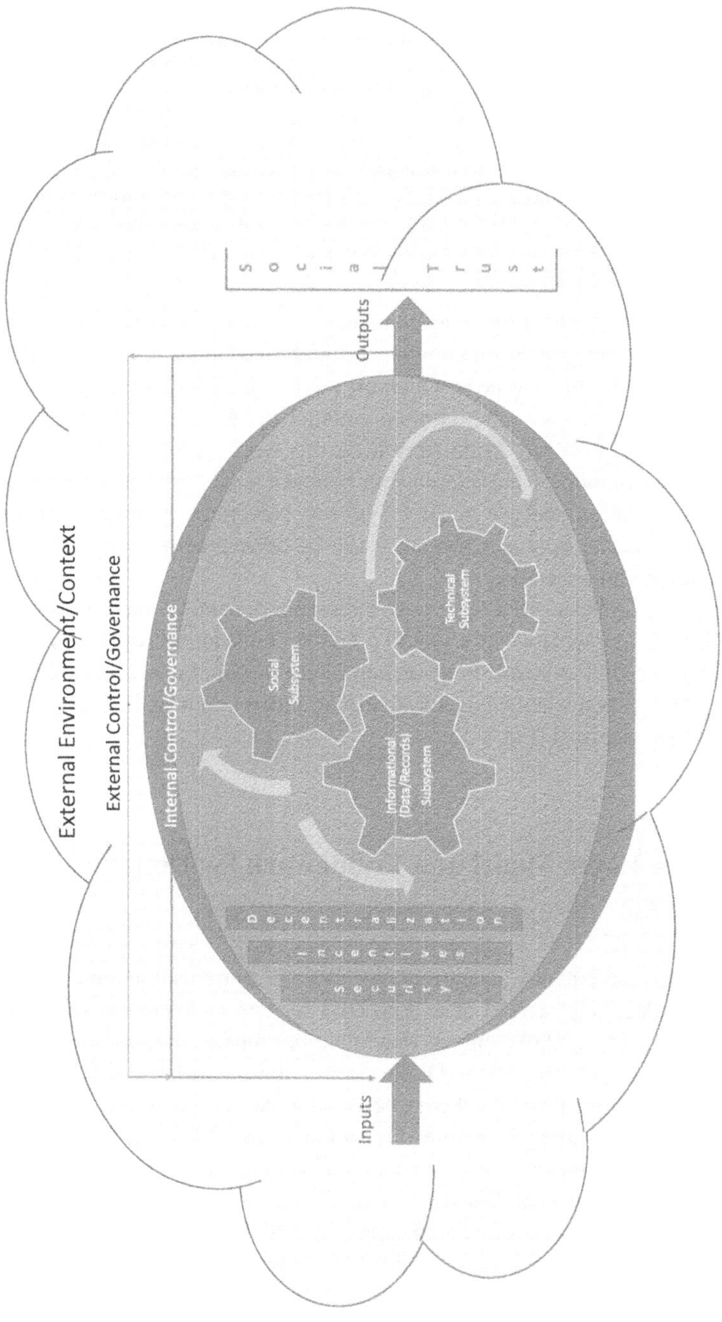

Fig. 7.2 Revised "three layer" blockchain/DLT system model

certain properties of DLT systems. Finally, though not initially identified as a separate theme, our explorations highlight the importance of recognizing that DLT systems exhibit temporality, as do their properties. In the remainder of this section, we discuss the five themes addressed in the preceding chapters, plus the element of temporality, fleshing out how they interact and might interact in DLT system dynamics.

7.4.1 Governance

Governance operates internally and externally to the system. External governance is usually a social form of governance, at the present time; that is, it is governance by social actors and social institutions (e.g., the "rule of law"). Internal governance, on the other hand, results from in-built mechanisms of self-regulation, which are often portrayed as technical, i.e., the consensus/incentive mechanisms of DLT systems, but which we argue involve more complex interactions among all three sub-systems/layers as described by Walch (2018). Following De Filippi and Wright (2018, p. 208), there are four key mechanisms by which to instantiate governance and that directly or indirectly regulate DLT systems: *laws, code, market forces, and social norms*. These mechanisms may be found and configured, to varying degrees, across the three layers.

Social norms function as a logic that determines and constrains the operation of both external and internal governance mechanisms of a DLT system. For example, as we point out in Chap. 2, "with developers, there may be norms determining that a developer will not try to thwart the system or that contributing to an open source software project is a virtuous act of contributing to the common good. Similarly, there may be norms around reputation—if a developer is seen to be trying to harm the system or seen to be incompetent, such behaviours will damage their reputation and future earnings."

In high trust and stable social environments, the goal of social trust in the design and operation of DLT systems may be achieved via internal control mechanisms (i.e., they may operate "trustlessly" or in a self-regulating manner) *if* such mechanisms do not fundamentally conflict with external control mechanisms (e.g., societal laws and regulations, organizational norms) in effect within the broader environment in which the DLT system functions. We note that these external control mechanisms may, themselves, be in conflict in cases where the environment in which a DLT system operates is relatively heterogeneous (e.g., when it operates across geopolitical jurisdictions, therefore encountering conflicts of laws or "logics" motivating social actors' actions). Such conflict with the external environment introduces friction that might prevent the DLT system from operating effectively, i.e., achieving homeostasis, or that destabilizes it. The avoidance of these tensions might explain why it is more common, and easier, to rely upon traditional mechanisms of control, i.e., the exogenous ones, than to rely principally upon a self-regulating system model of DLT systems. Moreover, because the internal control mechanisms are predominantly conceptualized as technical and the external control mechanisms as socio-political, governance is complicated by the absence of a well-defined socio-technical interface that translates between the two types of governance mechanisms. Neither type of governance is well understood by those social actors involved in the design of the other type of governance, e.g., lawyers in the case of the social layer, or software developers in the case of the technical layer. More work is needed to create a framework for interaction in relation to system governance, given that DLT systems seem to have to rely upon both types of governance in their operation.

Governance is an adjustable DLT system design element in that designers can determine to what extent the system will be self-regulating (i.e., rely upon internal control versus external control). Equally, in considering the internal control mechanisms, designers make different choices about the extent to which the internal control mechanisms will rely on social trust among system users and stakeholders, the trustworthiness of the ledger, or technical mechanisms for establishing trust. For example, in the context of public permissionless blockchains, system designers prefer to rely upon self-regulation, making an assumption that there is no inherent basis by which trust can be established among interacting social actors (i.e., users/participants in the system), that social actors must be able to place a high degree of trust in the ledger as evidence of facts about acts, and must rely upon technical algorithms (e.g., Proof of Work) to assure the trustworthiness of the ledger. In private, permissionless DLT designs, these assumptions are relaxed, or at least differ, with more trust being placed in social actors (by reliance on contractual terms and conditions and the fail-safe of external control by rule of law). Similarly, although the ledger must still be deemed to be trustworthy, it is only to a certain level of assurance with ultimate resort to contractual terms and conditions should internal controls prove ineffective, which allows for less rigorous technical mechanisms of assuring the trustworthiness of the ledger (e.g., use of Practical Byzantine Fault Tolerance rather than Proof of Work consensus). In essence, then, designers can choose how the DLT sub-systems and internal and external controls, or governance mechanisms, will be designed and interact to achieve social trust. Caution is urged against assuming too singular a reliance on either self-regulation or external regulation. Instead, as Kauffman (2013) argues the case for "[w]ise enablement via laws and regulations, [the] cascading consequences [of which] we cannot foresee." (p. 21).

7.4.2 *Incentives*

An incentive can be characterized as a type of "actuator" that operates on the "logic" or set of instructions for processing information possessed of all system components. An incentive triggers the operation or actions of each component and affects the operation of the system as a whole. As such, incentives can be characterized as forming part of a system's internal control sub-system or self-regulation mechanism, but also as influencing its external governance.

Incentives can be designed to alter the behavior of an actor or component as it contributes to the operation of the system as a whole. The most common example of this is the use of cryptocurrency block rewards to incentivize miners to undertake the proof of work needed to confirm blocks in some blockchains. Cryptoeconomics is an emerging field of study focused on the design of incentives and mechanisms in a blockchain network (Voshmgir and Zargham 2019). The design of incentives and rules of behaviour (mechanisms), generally referred to as "mechanism design", may affect the operation of components in the same sub-system (e.g., consensus among nodes within the technical sub-system in the case of Practical Byzantine Fault Tolerance) or may also operate across sub-systems (e.g., cryptocurrency block rewards that incentive social actors to operate technical components to confirm blocks).

Incentive mechanism design relies upon making assumptions about the "logic" of system components, e.g., miners, and their underlying motivations. For example, it is typically assumed that miners are acting as rational economic actors, but we note that this assumption may be false. Incentive mechanism design is complicated by the fact that components participating in the operation of a particular DLT system may, in fact, each belong to independent and distinct ecosystems. These ecosystems may operate according to norms and practices that influence the way that components (e.g., social actors) process information and may also introduce incentives that compete with or confound the incentive mechanisms designed into DLT systems. This may cause the components of a DLT system not to behave in accordance with designers' or other system components' expectations, which can, in turn, prevent a DLT system from achieving its goal (i.e., social trust). We agree with Zhang et al. (2020) that designers may have, at best, indirect control over the incentive structure with little control over the exact behavior of actors or actants in the context of designing DLT systems.

A key challenge relating to incentive mechanism design in DLT systems concerns adaptation, particularly where incentives have been built or "hard coded" into the self-regulating internal control mechanisms ["rule of code" (De Filippi and Wright 2018)]. As pointed out in Chap. 3, changing the original protocol or code requires consensus from the participants in the network and there may be no clear agreement on how to achieve consensus on these kinds of changes. In other words, while the Bitcoin mechanism provides an incentive compatible framework for agents to carry out normal transactions on the network, there is no mechanism that allows for necessary periodic updates to the protocol itself. This has led to schisms in DLT networks, commonly referred to as "forks" and, as described in a recent paper (Noda et al. 2019), which showed that miners may prefer high Bitcoin price volatility to maintain the value of their mining "application-specific integrated circuits" (ASICs) and therefore, may refuse proposals and innovations for stabilizing the price of Bitcoin. Thus, hard coded incentives can prevent necessary adaptive adjustments to the operation of a DLT system. How to allow for evolutionary adaptability in DLT systems remains an open challenge.

Mechanism design posits that if incentives are set appropriately it is possible to achieve desirable outcomes and, conversely, when not set appropriately, the outcomes can be negative. However, an open question is whether mechanism design is, well, too mechanistic for complex systems. Some systems theorists argue that systems do not operate strictly by rules of cause and effect. Rather, they may be self-organizing. As Kauffman (1995) points out:

> [t]he past three centuries of science have been predominantly reductionist, attempting to break complex systems into simple parts, and those parts, in turn, into simpler parts. The reductionist program has been spectacularly successful, and will continue to be so. But it has often left a vacuum: How do we use the information gleaned about the parts to build up a theory of the whole? The deep difficulty here lies in the fact that the complex whole may exhibit properties that are not readily explained by understanding the parts. The complex whole, in a completely non-mystical sense, can often exhibit collective properties, 'emergent' features that are lawful in their own right. (p. i–ii).

Kauffman (1995) refers to this phenomenon as "order for free". An open question, then, concerns the limits of mechanism design, i.e., is it possible to know all of the emergent capabilities of a given system, and can DLT systems also achieve order for free?

7.4.3 Security

Data integrity, in many cases, is essential for social trust. As an example, until the double-spending problem was solved by Bitcoin, people could not have confidence that their digital currency transactions would not be manipulatable (i.e., double spent). Once the double spending problem was solved using a peer-to-peer distributed timestamp server to generate cryptographic proof of the chronological order of transactions, the goal of social trust in relation to the use of digital currencies became achievable. Thus, the property of security, and more specifically, integrity in the CIA (meaning confidentiality, integrity and availability) security triad (see, for example, Stewart et al. 2012), an essential but non-functional property of a DLT system, is mediated through the integrity (or immutability as it is sometimes referred to) of a distributed ledger. This security, which is achieved via the operation of technical components to produce data integrity, provides a foundation for social trust. These relationships hold true for all effective applications of distributed ledger technology, including blockchains.

Given the above, any threats to the integrity of data, such as the well-known "51% attack", are particularly dangerous in the context of DLT systems. But a DLT system's strength in protecting data integrity can also be its weakness. As mentioned in Chap. 4, "the open aspect of some platforms that accept smart contracts can allow for malicious code to be introduced and executed. In this case, the immutability of the code on the blockchain can be problematic. Introduced vulnerabilities cannot be fixed as smart contracts are immutable by design. Code audits therefore become of utmost importance prior to deployment."

There is an important relationship between security, in particular, and the maintenance of data integrity, in that social actors may either be incentivized or disincentivized to maintain the security of a DLT system. To illustrate this point, consider that if Bitcoin miners are rewarded in Bitcoin for confirming blocks then presumably, at least in theory, there is a disincentive to mount an attack on the integrity of the network that would lead to a reduction in the price of Bitcoin, which would surely be the case if people were no longer able to rely upon the fact that Bitcoin protected against the double-spending problem. Of course, sometimes, these assumptions about the motivations of network participants do not hold true; there are other motivations and incentives coming from competing ecosystems of which social actors may form a part. Thus, the assumptions underpinning security models must be carefully examined and constantly reviewed.

Confidentiality, whether it be of data/records or transaction origins, and availability are still important concerns, but in general in the context of DLT systems matter less than protecting integrity. The relative weighting given to these properties is, of course, dependent upon the particular use case. Tensions between these security properties can arise, as in the case of the need to remove personally identifiable information from distributed ledgers in compliance with privacy laws and regulations such as the EU's General Data Protection Regulation (GDPR), which would conflict with preserving the integrity of a distributed ledger (Hofman et al. 2019). No easy resolution to the tension between data integrity and data confidentiality (or, more specifically, privacy) has yet been found.

In addition, taking into consideration the extended CIAAA (referring to confidentiality, integrity, availability, accountability, and audit) security framework (Iguer et al. 2014), authenticity of data/records is only partially satisfied when data integrity is achieved (Lemieux 2017), even though it is a necessary pre-condition to it. Yet, Lemieux (2016, 2017) and Bui et al. (2020) also point out that bit-wise data integrity does not satisfy the requirements for records integrity, since records must remain accessible and human interpretable over time, which often requires many micro-changes to the bit structure of data, or data transforms, in order to render the data readable after successive software and system upgrades/changes. Thus, a different standard of integrity must apply to records than to data. Rather than bit-wise integrity, records must have consequential integrity; that is, they must retain all of the original elements of content, intellectual form, and physical structure necessary to continue to serve as evidence of and, as necessary, to maintain the state change (e.g., a transfer or maintenance of rights) they were originally created to produce. There is, as yet, no definitive measure of the acceptable bounds of changes to records integrity despite research aimed at identifying the "significant properties" of records that require preservation (Becker 2018; Yeo 2010). It may only be possible—since records are meant to create, extinguish, or maintain juridically relevant acts—for acceptable bounds of changes to records integrity to be determined over time within the social layer or sub-system (i.e., by historians interpreting facts about acts documented in records, or by courts with reference to the rule of law established in particular juridical systems).

Accountability within the CIAAA framework is dependent upon non-repudiation. In turn, non-repudiation, which aims to ensure that an individual or entity cannot deny the authenticity of their digital signatures, or that they were the originator of a particular message or transfer, depends upon authenticity (e.g., of digital signatures) and data integrity (i.e., that the message has not been tampered with).

Accountability also often is said to rely upon transparency (Fox 2007; Hood 2010), in the sense that the actions of an individual or entity to be held to account must be visible to the individual or entity to whom the other individual or entity is accountable. Thus, there must be an accounting, or accounts, which, in the world of analogue recordkeeping typically took the form of ledgers and other forms of records, but now may take the form of distributed ledgers implemented as DLT systems. Thus, the transparency of a DLT ledger, achieved by virtue of all participating nodes contributing to and maintaining a full copy of the ledger (hypothetically speaking), is an important aspect of achieving accountability in the CIAAA security framework.

We note that transparency is not a static property of DLT systems but is configurable. For example, Hyperledger Fabric has introduced the concept of "channels" to maintain the confidentiality of certain transactions; users need to be subscribed to a specific channel to be able to view the transactions taking place within the bounds of that channel (Androulaki et al. 2018). In a similar vein, Monero has introduced Ring Confidential Transactions (RCTs) to hide the actual transaction amounts between a sender and recipient instead of relying solely on obfuscating the identity of the sender, as in other privacy preserving blockchains (e.g., Zcash) (Biryukov et al. 2019; Noether 2015). When transparency is constrained in these ways, so too may be accountability. Thus, design of DLT systems not only involves the commonly mentioned trade-off between security and scalability (see, for example, Gountia 2019) but also trade-offs among different security properties, i.e., confidentiality versus transparency, or data integrity vs. privacy. Equally, there may be trade-offs within the same security property as applied at different layers or levels of abstraction within the same layer, e.g., bit-wise data integrity may be in tension with the consequential integrity needed for records.

Bitcoin emerged as a system governed by the rule of code, i.e., a cryptographic proof that established the chronological order of transactions. All nodes on the network, in theory at least, contributed to undertaking transactions, confirming blocks, and maintaining the network. In this sense, Bitcoin does not have "users" of the network; rather, it has participants. Moreover, in the early days, the individuals involved in this work were highly "code" literate, and thus can be seen as experts. As such, Bitcoin was designed by and for experts, and can be characterized as an expert system in the broadest, non-AI meaning of the term. Over time, however, the theoretical vision of Bitcoin network operation and the lived experience of operating the network have diverged, with mining pools and the community influence of some participants taking on greater power.

In other DLTs, such as IOTA (Lamtzidis and Gialelis 2018), specialized nodes have emerged, which have greater power and specialized functions, while in other DLTs, such as the permissioned DLT Hyperledger Fabric (Androulaki et al. 2018), the differentiation between system admins and users has been clearly designed into the system architecture. At the same time, as many DLT networks have expanded—which they needed to do to increase data integrity and achieve the system goal of social trust and thereby homeostasis (i.e., the greater the number of nodes, the greater difficulty in tampering with the history of the chain)—a greater number of individuals with less expertise have become participants. Not having the same level of expertise, they have struggled with the management of private keys and with other aspects of full participation in the operation and maintenance of DLT networks. This, in turn, has created the problem of "useable security" in DLT systems which, from this perspective, can be viewed as a problem necessitating, in some cases, recentralization of decentralized power in DLT systems. The problem arises because such systems, as they are currently designed and operated, are simply too complex and require too much expertise for some participants to use securely. As a result, some participants have taken on more of the classic role of "users", delegating certain responsibilities to others in order to abstract away the complexities of full participation. Maintaining cryptocurrency on exchanges and relying on browser-based cloud wallets are just two examples of this strategy. In doing so, these participants, having converted themselves into users, have given away some of the power (and responsibility) that they held as fully participating nodes in a peer-to-peer decentralized system. New centralized entities in the wider DLT system, such as exchanges and software wallet providers, have emerged. In turn, this has made these users susceptible to security threats such as thefts from cryptocurrency exchanges or software wallet hacks (Boireau 2018; Kim and Lee 2018). To address these security challenges, we must successfully apply traditional security techniques designed for centralized systems to the new entities to which decentralized ecosystems have given rise, develop new security techniques for decentralized systems that do not involve recentralization, or enable co-evolution of the expertise of system participants. The most effective mix of these techniques will, again, likely depend upon the context of the particular DLT use case.

7.4.4 Decentralization

The architecture of each sub-system/layer of a DLT system can be centralized or decentralized to varying degrees to affect the operation of the system as a whole.

A decentralized technical architecture *may* facilitate a decentralized social architecture (e.g., decentralized social interactions), but not necessarily. The two can operate independently. Equally, a *desire* among social actors for decentralized social interactions does not necessarily yield a decentralized technical architecture. In order for social decentralization to occur, social actors must have the incentive and the means to exit (Hirschman 1970; Markey-Towler 2018) existing juridical-legal, political, social, and environmental system constraints. In reality, this rarely happens, which is why it is so challenging to implement (if not design) fully decentralized systems at all three layers. Indeed, the technical decentralization aspects of DLTs may create barriers to their own adoption, unless carefully designed and implemented, due to inherent conflicts with existing, often centralized, social architectures.

The most likely environment to yield the conditions to enable full decentralization are collapsed states, or ones in which existing institutional structures are very weak or not operating effectively, such as countries experiencing economic or political collapse, or weakness for endogenous (e.g., corrupt elites) or exogenous reasons (e.g., public health crises), or constrained environments that operate as quasi-socio-institutional green field sites (e.g., refugee camps). Normally operating, highly institutionalized settings, such as tightly regulated sectors or markets, are less likely to allow full decentralization until such time as the environmental factors evolve to align more fully to the decentralizing objectives of DLT systems (i.e., when decentralized social organization is more trusted than centralized ones and centralized entrenched power structures have disappeared or their grasp on power is such that decentralization can effectively take root.)

The relationship between decentralization and social trust is indeterminate, socially embedded, and mediated through power. On the one hand, in the context of public permissionless DLTs, there is a working assumption—similar to that expressed in such texts as May's 1992 Crypto Anarchist Manifesto—that technology can achieve decentralization of social architectures, and concomitantly that decentralization of social power to all participants in a network leads to greater social trust, since all actors can depend upon the operation of the DLT system to fairly and dispassionately regulate their interactions and behaviors. This working assumption may not hold true.

There is a relationship between governance and decentralization that involves the distribution of power among network participants, but we do not yet fully understand the nature of that relationship. As Walch (2019b, p. 40) explains: the term 'decentralized' is generally used to describe how power operates in blockchain systems—suggesting that power exercised by people in these systems is diffuse rather than concentrated. This is critically important, as our understanding of how power is exercised within these systems will shape conclusions about how responsibility, accountability, and risk should work for them.

Decentralization at any layer may be explicitly designed to overturn and diffuse existing power structures, and thus has the potential to impact upon all legal decisions surrounding DLT systems and existing power structures (e.g., the "rule of law"). The working assumption of public, permissionless DLTs tends towards the view that centralized social architectures lead to arbitrary definition and application of the "rule of law" in a manner designed to ensure that the central actor's grasp on power is retained and even enhanced to the detriment of other actors in the system. To some degree, democratic political theories also adopt this stance, in that democracy posits and incorporates mechanisms to give "power to the people" to counter these centralizing tendencies. Nevertheless, to achieve social coordination and action, even in a democracy, people must rely upon centralized government agencies/actors to which they delegate the power to make decisions on their behalf, e.g., approximately every 4 years by means of elections, and solve collective action problems. This can lead to information asymmetries, which are imperfectly resolved by such public accountability mechanisms as anti-corruption agencies and freedom of information laws which, similar to DLTs, rely upon transparency of records and information to redress the power imbalances caused by centralization and the introduction of attendant information asymmetries in the system.

The public, permissionless DLT conceptualization of decentralization vis-à-vis social trust and power bears a fundamentally negative connotation of centralized social power and social architectures. On the other hand, some (e.g., Atzori 2015;

Lemieux 2019) see dangers in social decentralization potentially made possible via DLTs, pointing out that they do not necessarily ensure a fair distribution of power or protection of individual rights, and thus could lead to a reduction in social trust. Moreover, as we point out in Chap. 5, centralization may be invoked even in decentralized systems as a necessary means to address failures in the operation of the system and associated internal control mechanisms. This argues against an overly binary approach to centralization and decentralized systems (i.e., decentralization "good" vs. centralization "bad") vis-à-vis the effective operation of the system and the overarching goal of social trust. Rather, it argues for a more flexible and dynamic—what may be characterized as "oscillating"—view of decentralization in relation to centralization in DLT systems. Another open question in relation to decentralization and power remains whether, even if all sub-systems are highly decentralized, they will operate in a manner leading to social transformation outside of the DLT system, e.g., decentralization of political power. To illustrate, as we point out in Chap. 5, "collusion and faction building between nodes can influence willingness to sign, and even security issues come to bear—were a denial of service (DoS) attack mounted upon sufficient nodes to prevent their interaction with the network, the remainder of nodes, if colluding, could agree to admit a disingenuous block to the chain—a so-called '51% attack'." Thus, to make assertions about the relationship between the design of decentralized sub-system components and the effects of those designs requires much more theoretical and empirical research on a social theory of change in relation to DLT systems.

7.4.5 Provenance

The ability to trace the provenance of a cryptoasset, possibly representing a real-world asset or a native digital asset, is a functional capability rather than an inherent property of a DLT system, its sub-systems, or components, and is discussed in much greater detail in Chap. 6. Yet, this capability leverages the peer-to-peer distributed timestamping and ability to cryptographically prove the chronological order of transactions afforded by DLT systems. Clearly, not all DLTs trace the provenance of cryptoassets and, in some cases—as when individuals "wash" tainted coins—there may be a desire to avoid this capability altogether. So, provenance tracking is a design choice rather than an inherent property of all DLT systems. Nevertheless, in many cases the provision of a trustworthy technical system for tracing the origins, ownership, and authenticity of cryptoassets can enhance social trust, as discussed in Chap. 6. The social trust brought about by provenance tracking is mediated through transparency (i.e., knowledge about the origins of something or someone that gives confidence in that person or thing), and accountability. This is, in turn, mediated through a distributed ledger (within the data/records sub-system), the integrity of which must be capable of being relied upon.

7.4.6 Temporality

We often design as though we live in the eternal now, and as though the properties and capabilities of our systems are predictable and permanent. DLT systems, however, are not static and their properties and capabilities may not be entirely predictable. Through recursive temporal configurations, for example, elements of design such as decentralization may be established via technical means (e.g., through protocol design), change over time as people make decisions (e.g., the social layer) about what transactions should be sanctioned on the blockchain (e.g., the data layer), and may generate new configurations at the data and technical layer that result in recursive configuration of the social layer (e.g., splits of blockchain social communities surrounding technical forks). In systems theory, small changes that have larger effects across an entire system are referred to as the "Butterfly Effect" (Lorenz 1995).

Like other open systems, degradation, disorganization, and decay of, and within, DLT systems can be expected (Dekker 2016). One example of this is likely degradation in the accessibility of DLT evidence, or proofs, of facts and acts over time due to technological obsolescence (e.g., deprecated software protocols, outdated hardware, etc.), loss of semantic meaning due to missing or failed linkages between ledger records and the context of their creation [i.e., absence of the archival bond (Lemieux and Sporny 2017)], and ecosystem governance failures leading to dissolution of DLT networks. This may lead to loss of critical information, such as identity records, land titles, or medical records. Thus, the possibility of entropy over time needs to be considered in the design of DLT systems. More research is needed in this area, and the general theory of entropic DLT decay could help to understand, and address, the entropy processes associated with DLTs over time.

DLTs, being open systems, will be affected and need to respond to changes in the external environment in which they operate, i.e., they will need to adapt. As a case in point, as we discuss in Chap. 4:

> Longevity of security is critical in blockchain security design and implementation to ensure sustainability of the blockchain and its data in the long run. Future threats to the underlying security mechanisms, such as the quantum threat to standardized cryptography and technological obsolescence of blockchain software, and their long-term implications on longevity of blockchains, should be considered and planned for now. The challenge here is to design a system that is going to resist all future attacks and the creative ways adversaries are going to use to try to undermine the security of blockchain systems, as well as be secure against the exigencies of time. This is a near impossible task. Instead, a more practical approach would be to plan and design for agility so that we can switch between algorithms when a new one is necessary, or migrate seamlessly and without disruption to new software protocols.

An important question for further exploration, then, is the one asked in Chap. 4: Can DLTs as complex systems adapt or be designed to be adaptable? As that chapter suggests, we may be able to look to nature for strategies that other adaptable highly complex multi-agent systems use to deal with complexity, such as distribution and redundancy, adaptation to change, limited life expectancy, resilience in the face of disturbances, and evolution.

7.5 Using the Three Layer Model in the Design of DLT Systems

Building upon the conceptualization of DLT systems as complex systems, and upon our increased depth of understanding about system elements and their interrelationships, we extend this thinking further by drawing upon complex systems theory to propose a question-led framework accompanied by a suggested approach to developing mathematical formalizations to enhance the toolkit for blockchain system designers.

Each question (see Table 7.1) in our framework aims to elicit essential information required to design a DLT system to achieve a specific social trust goal, though we make no strong assertions at this point that the questions are comprehensive. In addition, we note that a variety of different methodologies may be used to answer these questions. In this volume, contributors have suggested approaches as diverse as grounded theory, archival diplomatics, human-centred design, mechanism design, algorithmic game theory, ecological interface design and mathematical modeling. The advantages and disadvantages of these approaches vis-à-vis exploring and answering the questions in our model deserve further treatment and remain as future work. We adhere to the view that design of DLT systems, as complex systems, benefits from a multidisciplinary approach. As Wade et al. (2019) observe, there is still a need to identify a "common language to communicate among practitioners with different training" with a view to generating designs:

- To create a methodology that is logical, easy to follow, and can be taught as part of any curriculum that teaches approaches to problem-solving, complex systems, and product development.
- To create a process that promotes exploration of innovative ideas and a systematic way to select the most promising option (engineering design).
- To promote a process that results in a low-risk, realizable, and profitable solution (systems engineering).
- To create resulting products that satisfy a real user need (design thinking).
- To demonstrate that resulting products are sustainable in the changing and increasingly connected marketplace (systems thinking).
- To verify that the approach can be used for systems which are humancentric, including complex social systems (systemic design).
- To establish a framework that can be tailored for use on systems and products of all types (agile systems/software engineering).

They see value in an iterative dialogic approach that engages broad disciplinary and epistemological approaches—humanism, social sciences, and engineering—reflecting that:

> Through system and design thinking, critical assumptions must [be] identified, and anthropological approaches should be used to establish ground truth. These can be short studies, or they can involve weeks or even months of fieldwork observation, which might be very difficult for action-oriented engineers to conduct, particularly where observation skills are critical. In these situations, knowledge in the development of design of experiments is critical, and scientific approaches of discovery are highly valued. Ideally, this is where engineers might enlist the assistance of those who are experts in the field, particularly in the social sciences. (Wade et al. 2019, p. 8)

Table 7.1 Question-led DLT system design framework

	System Goal: What is the stated purpose of the DLT system? How does the stated purpose support social trust? What problem(s) should this system solve? What use cases is the system designed to support? **System Constraints**: What behaviors must the system be designed not to tolerate? What is the system's space of permissible actions? **System Capabilities**: What capabilities must the system possess in order to achieve its goal within prescribed constraints?			
	Social Sub-System	**Data/Records Sub-System**	**Technical Sub-System**	**Governance Sub-System**
Environment: Is the environment in which the DLT system operates relatively homogeneous, or is it more heterogenous? What assumptions about the environment does the DLT system make in its design/operation? What aspects of the environment does the system rely upon? At what points and for what purposes are these relied upon? How aligned with all aspects of the environment is the DLT system? What elements from the environment influence or constrain designers of the DLT system? What elements from the	Who are the social actors in the DLT? How are they identified/represented? How are their identities regulated? How does the DLT system empower or constrain their agency? What types of actions of social actors are forbidden, encouraged, or tolerated? Where is power located among social actors? What values are important to the social actors in this system? What expectations do we have of the behaviour of the social actors? What actions will or might they take? How are these actions expected to impact upon others?	How does the ledger serve to support social trust in the context of the DLT system? What data is captured/flows through the system to support the system goal? What records are generated to support the system goal, either on ledger or off ledger? What data is captured/flows through the system to support the interaction of the actants? What records are generated to support the interaction of the actants, either on ledger or off ledger? How are the data/records actants in the system identified and how are their identities regulated? What data and/or records must	What are technical actants in the DLT system (e.g., sensors, vehicles)? How are the technical actants in the system identified and how are their identities regulated? How do the technical actants serve to support social trust in the context of the DLT system? What capabilities and properties do they require to support the system goal? What is the system architecture? What is the network architecture/topography? What social actors control the technical actants in the DLT system? How do these social actors empower or constrain	How much reliance will there be on internal or self-regulating governance vs. external governance under normal operating conditions? Under abnormal operating conditions? How will consensus decisions be made among technical, data/records and social actants/actors? What incentives are or will need to be put in place so that the consensus mechanism operates in a manner that supports the goal of the system? How should decision management rights and decision control rights be allocated

environment influence or constraint system actors or actants?	When is the consent, permission, and authority of social actors needed, granted, or assumed? Will some social actors act on behalf of others? On what (moral, legal?) ground do they implement the will of others? Which others? How do social actors need to exercise, or do they exercise discretion when conflict arises?	the system store? (What are the legal or regulatory obligations?) What data and/or records must not be stored in the system? (For purposes of privacy, financial risk management, or corporate policy.) Are there data and/or records that require special consideration? For example, are there data and/ or records containing personally identifiable information that require special treatment under law? Are there data and/or records that must not be kept indefinitely? Where are records stored? How are they propagated across networks? How are the intellectual components of the record assembled?	the activity of technical actants? What level of authority/authorization do the technical actants have?	among various interacting components (whether social, data/records, or technical)? How will disagreement about decisions be resolved?
	Temporality: What known future changes will the system have to be able to respond to? What mechanisms need to be put in place to assure the longevity of the system? Could future events bring about consequences whereby the platform ought to be completely replaced or cease operation? How will the governance sub-system address actors'/actants' changing relationships to the system over time? How will risk factors be addressed, including those that lie unknown in the future and that may present existential or systematic risk? How has/does power shift among social actors over time?			

Following a systems analysis and design-oriented approach, our framework begins with a consideration of the system's goals, constraints, and required capabilities. To illustrate, let us assume a system, as in the case of Bitcoin, with a *goal* of increasing social trust between interacting economic actors through disintermediating centralized banking institutions seen to be increasing their share of value undeservedly. To achieve the goal of increasing social trust, the system must operate in a way that keeps track of valid transactions between actors without the need of a central coordinating banking institution and must be *constrained* in its operations so as to prevent fraudulent manipulation of those transactions (i.e., it must prevent "double-spending"). In order to achieve its goal and operate within prescribed constraints, the system will need to possess certain *capabilities*; for example, it must record the transfer of a digital asset (e.g., a Bitcoin) from one rights holder to another, it must keep records of transactions in a ledger that all participants in the system can trust, etc.

The operation of all open systems is shaped, to varying degrees, by their operating environments. Thus, consideration of the external environment, or context, of a DLT system (i.e., those elements outside the boundary of the system or its space of permissible actions) is recommended, especially since the system's external environment may not be homogeneous as when systems operate across geo-political jurisdictions or different social actants are members of very different ecosystems (e.g., system users versus core developers). For all sub-systems, it is important to think about what aspects of the environment the system will rely upon for its operation, and at what points and for what purposes these aspects are/will be relied upon. This is particularly important with respect to the governance sub-system, since self-regulation and external regulation may have a strong tie and, in some cases, DLT systems will rely almost entirely upon external controls as their governance mechanism. As was discussed in the previous section, it is also important to consider how well-aligned the DLT governance sub-system is with all aspects of the governance environment, since misalignment will cause friction and potential operating or governance problems for the system. More broadly, consideration should be given to how norms, values, motivations, and incentives from the environment influence or constrain operation of the DLT system and the actions of social actors comprising the system. Reflexively, designers should consider how their own norms, values, motivations and incentives act as influences or constraints on the system, since some case studies have shown that designers' own economic interests may interfere with realization of fully decentralized system designs (Lemieux et al. 2020). Designers also should examine what assumptions about the environment they are making in the design/operation of the DLT system to avoid designs that embed false assumptions about the behaviour of social actors, accuracy of data, or availability of underlying technical systems.

Within the social sub-system, the social actors engaged in the operation of the system will need to be identified, and consideration must be given to how their identities will be instantiated and regulated by the system. In the Bitcoin example, for instance, social actors control addresses. In other systems, they may control other actants with DLT system accounts, e.g., smart contracts or Internet of Things (IoT) devices. In considering the social actors, it is important to think about how the DLT system is intended to empower or constrain the agency of these actors e.g., what types of actions are forbidden, encouraged, or tolerated of social actors? As a foundation to understanding the system's potential effect on agency, it is necessary to think about what values are important to the social actors in this system and what norms or other frameworks motivate or constrain them. Resulting from this understanding, consideration should be given to what expectations we might have of the behaviour of the system's social actors, and what actions they will or might take and how these actions are expected to impact upon others. In addition, since DLT systems often operate to achieve social trust by redistributing power among social actors, it is important to assess where power is located among social actors, or where it is intended to be located, and how other social actors, such as those within the DLT system's operating environment might react to any redistribution of power. In addition, it is important to consider how social actors exercise discretion when conflict arises and when their consent, permission, and authority is needed, granted, or assumed.

DLT systems, according to our theory, have an information (data/records) sub-system that also must be explicitly considered. This is described in DLT systems as a "distributed ledger", even though in many DLT systems this sub-system may perform many recordkeeping functions well-beyond the traditional function of a ledger (e.g., in DLT systems supporting land transfers, the data/records sub-system may serve as a registry). An initial question to ask is how the ledger serves to support social trust in the context of the DLT system. It is also necessary to understand what data/records are created, captured and flow through the system to support the system goal, and where and how data/records are created, captured and stored. A primary reason to ensure an understanding of data/records and their flows and storage is in connection with considerations of privacy regulation, financial risk management, and corporate governance. Depending on the type of data/records, there may be special considerations for their creation, capture, transmission, location, and method of storage and retention, often arising from the external environment (i.e., juridical-legal system). Similar to social actors, the identity of record actants must be established and regulated as a foundation for establishing records authenticity. This might be achieved through the creation and instantiation of records metadata—such as, a description of the content of the record; the structure of the record; the business context in which the record was created or received and used; relationships with other records and other metadata; identifiers and other information needed to retrieve and present the record, such as format or storage information; and the business actions and events involving the record throughout its existence—which, if not present, may render the ledger less trustworthy as proof or evidence of the transactions recorded in the ledger. Arising from this analysis, DLT system designers will need to determine what records are or will need to be generated to support the system goal, either on-ledger or off-ledger.

Also to be considered is the technical sub-system, which includes identifying the technical actants in the DLT system (e.g., sensors, vehicles, applications, APIs, servers, networks, protocols, etc.). Similar to social actors and informational actants, consideration needs to be given to if and how physical actants' identities will be regulated. Designers will need to think about what technical capabilities are necessary to support the DLT system's goal and what components and properties of components are needed to achieve these capabilities. Representing these components in diagrams of a DLT architecture or stack (as discussed in Chap. 4) can be helpful in *thinking* about these elements as well as *communicating* about them. Importantly, designers should reflect on which social actors control the physical actants in the DLT system, and how these social actors empower or constrain the activity of physical actants. Thought must also be given to the incentives that drive the social actors who control the physical actants in the DLT system, and to their relative social power. Equally, attention must be paid to the relative processing power and energy consumption demanded of different components.

The governance sub-system, or internal control mechanisms, is a further aspect of DLT system design that must be considered. In particular, designers will need to think about how much reliance there will be on internal or self-regulating governance versus external governance under normal operating conditions, and what will happen under abnormal operating conditions (e.g., when the system's constraints are breached or the system is operating or being used outside the space of permissable actions). Consideration must be given to how consensus decisions made among technical, information, and social actants/actors (e.g., selection of consensus mechanisms) will be made, as well as what incentives are or will be needed so that the consensus mechanism operates in a manner that supports the goal of the system. As Zhang et al. (2020) note, "The role of the designer is to design a set of rules and incentives such that the system-level goal can still be achieved irrespective of the exact behavior of the agents." In governance systems, including internal governance systems, thought must be given to how decision management rights and decision control rights will be allocated among various interacting components (whether social, data/records or technical), and how disagreement about those decisions will be resolved (e.g., revert to external governance, special self-regulating procedures, etc.).

Finally, designers should give consideration to how the system will evolve over time. How will the governance sub-system, for example, need to address actors' changing relationships to the system, including how power may shift among social actors over time. Designers will need to think about what known future changes the system will have to be able to respond to, such as legal or regulatory changes in the external environment, and how the system and its actors will respond to these changes. There may be risk factors—including those that lie unknown in the future—that may present existential or systematic risk to the system, and there may even be future events that could bring about consequences whereby components need to be replaced to assure longevity or it may make sense for the platform to be completely replaced or cease to exist.

7.6 Towards a Mathematical Foundation for the Design of DLT Systems

In this chapter, we have suggested a model of DLT systems that extends our collective deliberations and emergent understanding about DLT systems, including blockchains, based upon general and complex systems theory. Modelling of such systems typically draws upon control theory—a subfield of mathematics that deals with the control of continuously-operating dynamic systems. In engineered processes and machines, the objective of control theory is to develop a control model for systems control using a control action in an optimal manner, without delay or overshoot, while ensuring control stability (Aström 2008). Control theory, in particular, has received considerable attention, as it represents a general methodology for creating adaptive systems, which, as we have noted, DLT systems need to be in order to achieve and maintain homeostasis.

To illustrate the approach, following Von Bertalanffy (1968), the five key themes we have discussed can be represented as a set of complex interacting (but not equivalent) elements, p_1, p_2, ..., p_n. Measures (including relative weights) of these elements can be represented by Q1 ... Qn. A collectivity of elements, e.g., sub-systems in a DLT system, can be represented as a multidimensional vector field wherein each cell in the field signifies measures of relevant elements within a sub-system that is both represented by, and comprised of, a set of differential equations, as shown in Fig. 7.3. This is essentially a state-space representation as, for example, is discussed in Zhang et al. (2020). Typically, a state-space representation, as defined in control theory, would be a mathematical model of a physical system as a set of input, output, and state variables related by first-order differential equations. Our model extends this approach to consider all three ontological layers or sub-systems—the social, data/records, and technical. This formalization recognizes that, in complex systems, both the *components* and the *relationships* between components are important.

The directional edges between each cell (vertex) represent the dynamic and highly interconnected nature of the system, with measures within each equation iteratively and recursively interacting with the measures in the other equations in *n* dimensional vector space. Systems also operate within environments, or contexts, which may, themselves, be signified as cells or vertices comprised of a set of differential equations with measures, including relative weights, of various elements.

In a state-space representation, state variables are variables whose values evolve through time in a way that depends on the values they have at any given time and also depends on the externally-imposed values of input variables [e.g., "the values of the states at a given time $t \in \mathbb{N}$ depend exclusively on the values of the states at time t - 1. Hence, given the initial values for the states at the origin of time (i.e., $t = 0$), it is possible to recover the states at any time $t > 0$ by solving this recursion" (Zhang et al. 2020, pp. 3–4)]. This formulation can model both linear and non-linear change, depending on whether the state at time t is a linear transformation of the state at time t - 1. While Zhang et al. (2020) use a linear state-space model, we recognize that

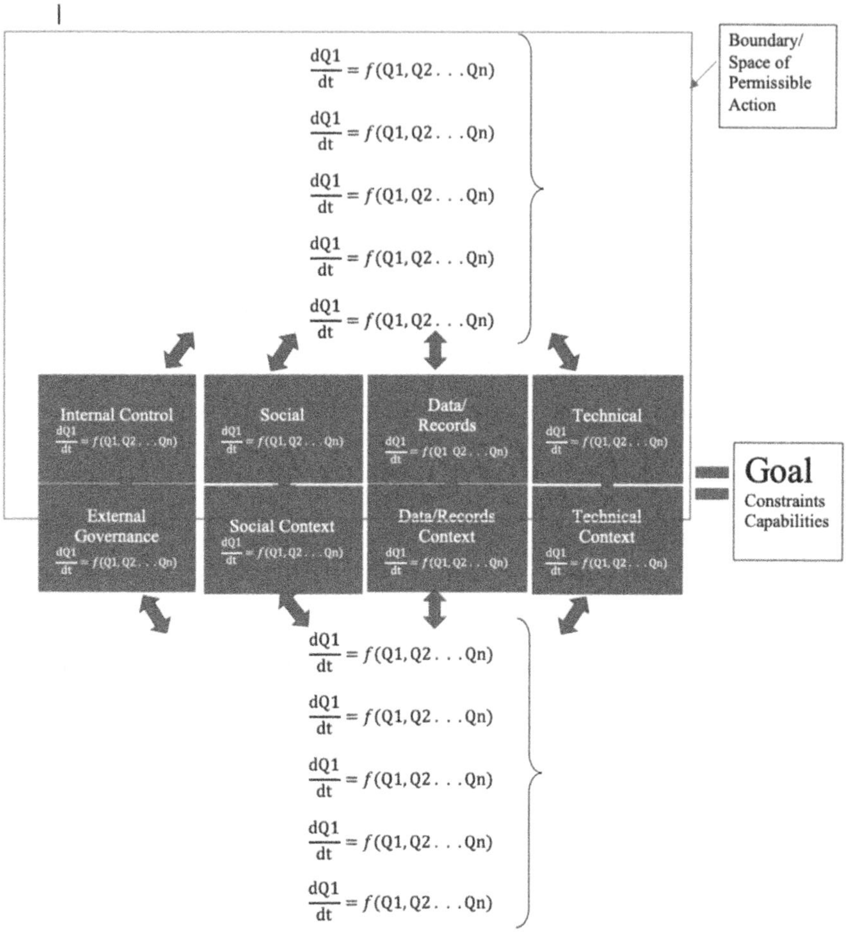

Fig. 7.3 Representation of a DLT system (and its context) as an *n* dimensional vector space

complex systems exhibit non-linearity. To capture this behavior, we propose to model a DLT system using a non-linear state-space model with probabilistic state transitions governed by technical designs and policies due to social or economic considerations. Such a stochastic non-linear modeling technique has been widely used in computer science and engineering for the analysis of complex networked systems, with a simple example given in Mitzenmacher's PhD thesis (1996).

One illustration of the application of control theory to DLT systems is Zhang et al.'s work (2020). It starts from a linear state-space representation for stochastic dynamical systems, where the state captures transaction addresses. It then introduces differential games in which each agent decides its own actions based on a set of rules and incentives. This modelling framework allows researchers to run Monte Carlo simulations to reason about the evolution of the system. This is particularly useful, since system designers can try different sets of rules and incentives to understand their pros and cons. On the other hand, this framework does not itself specify all the system parameters and their distributions. It simply states that "(d)etails around how to select and parameterize a distribution are beyond the scope of this paper." (Zhang et al. 2020, p. 3). As we will soon see, this drawback can be well addressed by taking an information-theoretic approach.

Filieri et al. (2015) do, however, observe that control-theoretical software implementations tend to be ad hoc and it can be difficult to understand and reason about the desired properties and behavior of the resulting adaptive software and its controller. Thus, we argue that while control theory is excellent for modelling DLT system dynamics from a top-down perspective, information theory may be much better suited to identifying underlying distributions for state-space representations and reasoning about the desired system-wide properties from a bottom-up perspective. Roughly speaking, an information-theoretical approach abstracts DLT systems into essential features and then develops mathematical models and theorems to understand several key performance metrics, such as transaction throughput (i.e., how many transactions per second a system can support) and confirmation latency (i.e., how long it takes for a transaction to be confirmed by all the participants). Such an approach often leads to fundamental insights and new engineering designs that have a potential to revolutionize existing systems (just as 5G is expected to be nearly 100 times faster than 4G).

One illustration of the application of information theory to DLT systems is Bagaria et al.'s work (2019). It has developed insightful mathematical models and theorems, which lead to a breakthrough design called Prism that achieves optimal transaction throughput and near-optimal confirmation latency up to the physical limit (such as the network capacity and the speed-of-light propagation delay). More importantly, the design of Prism solves the so-called "blockchain trilemma" popularized by Vitalik Buterin of Ethereum. The trilemma states that it is impossible to get decentralization, security, and scalability at the same time. In other words, no DLT system is decentralized, secure, and scalable. This trilemma comes from the real-world experience of trying to build DLT systems. For instance, Bitcoin and Ethereum are decentralized and secure but not scalable, while EOS and Ripple are secure and scalable but not decentralized. Is it possible to get all three at the same time? Prism answers this question in the affirmative by designing a particular DLT system based on ideas and tools from information theory.

After we see the application of both control theory and information theory, a natural next step is to combine these two in order to build a mathematical foundation for the design of DLT systems. We believe that such a combination is possible since control theory provides a top-down view and information theory gives a bottom-up view. More specifically, we can define a state-space representation based on models coming from information-theoretic approaches and then integrate such representation with optimal control theory for stochastic dynamical systems consisting of decentralized agents (with various social and economic goals). Here, we scratch the surface of a potentially rich and powerful mathematical foundation for DLT systems, which we believe will lead to a program of exciting new research.

7.7 Conclusion

In this chapter, we have synthesized and extended the results of a collective, multidisciplinary intellectual journey which began at the Peter Wall Institute for Advanced Studies' Workshop on *The Truth Machine: Exploring the Social, Records and Technical Potential and Pitfalls of Blockchain and Distributed Ledger Technologies*. At the outset of this journey, the interactions among social, records, and technical aspects (identified as the "three layers") of DLT systems were not sharply defined, but over the course of the workshop, and during workshop participants' collective chapter writing, the contours of the "three layers" and their relationships came into sharper focus. One contribution of this chapter has been to propose an integrative multidisciplinary ontological framework as a basis for synthesizing participants' emergent understanding of the layers and key themes of exploration. It has also extended the "three layer" model by drawing upon general and complex systems theory. Finally, it has proposed a framework to aid design of DLT systems.

We fully acknowledge that this extension of the "three layer" model of DLT systems as socio-informational-technical systems has a number of limitations, some due to unresolved logical inconsistencies, some to the need for further theoretical work, and some to the need for more empirical research. Some of the limitations are, however, inherent to all models, which are merely idealizations. As Korzybski (1933, p. 252) reminds us, "the map is not the actual territory." Thus, any model abstracts away from specific detail to allow for generalization. And, as Kauffman (2013, p. 29) writes, "We do not live in the world we thought in the Newtonian framework. We live in a world of unprestatable, new, unintended possibilities, opportunities, biosphere, econosphere, history. We do not know all the variables that will become relevant." Thus, there is always the danger that important assumptions or constraints are overlooked and that theoretical propositions prove to be false under certain conditions. No doubt this is the case with our model; however, in discovering its bounds and limitations, which we hope future research will uncover, our wish is that the model will serve as a starting point to develop a deeper understanding of DLT systems and how they might best be designed.

References

Ackoff, R. L. (1994). Systems thinking and thinking systems. *System Dynamics Review, 10*(2–3), 175–188. https://doi.org/10.1002/sdr.4260100206.

Androulaki, E., Barger, A., Bortnikov, V., Cachin, C., Christidis, K., De Caro, A., Enyeart, D., Ferris, C., Laventman, G., Manevich Singh, G., Smith, K., Sorniotti, A., Stathakopoulou, C., Vukolic, M., Weed Cocco, S., & Yellick, J. (2018). Hyperledger fabric: A distributed operating system for permissioned blockchains. In *EuroSys '18: Proceedings of the Thirteenth EuroSys Conference* (pp. 1–15). New York: ACM. https://doi.org/10.1145/3190508.3190538.

Ashby, W. R. (1991). Requisite variety and its implications for the control of complex systems. In G. J. Klir (Ed.), *Facets of systems science* (1st ed., pp. 405–417). Boston, MA: Springer.
Aström, K. J. (2008). Event based control. In A. Astolfi & L. Marconi (Eds.), *Analysis and design of nonlinear control systems: In honor of Alberto Isidori* (pp. 127–147). Berlin: Springer.
Atzori, M. (2015). *Blockchain technology and decentralized governance: Is the state still necessary?* https://doi.org/10.2139/ssrn.2709713
Bagaria, V., Kannan, S., Tse, D., Fanti, G., & Viswanath, P. (2019). Prism: Deconstructing the blockchain to approach physical limits. In *CCS '19: Proceedings of the 2019 ACM SIGSAC Conference on Computer and Communications Security* (pp. 585–602). New York: ACM. https://doi.org/10.1145/3319535.3363213.
Bailer-Jones, D. M. (2003). When scientific models represent. *International Studies in the Philosophy of Science, 17*(1), 59–74. https://doi.org/10.1080/02698590305238.
Beaudouin-Lafon, M. (2004). Designing interaction, not interfaces. In *AVI '04: Proceedings of the Working Conference on Advanced Visual Interfaces, May 2004* (pp. 15–22). New York: ACM. https://doi.org/10.1145/989863.989865.
Becker, C. (2018). Metaphors we work by: Reframing digital objects, significant properties, and the design of digital preservation systems. *Archivaria, 85*, 6–36. https://archivaria.ca/index.php/archivaria/article/view/13628
Biryukov, A., Feher, D., & Vitto, G. (2019). Privacy aspects and subliminal channels in Zcash. In CCS '19: *Proceedings of the 2019 ACM SIGSAC Conference on Computer and Communications Security* (pp. 1813–1830). New York: ACM. https://doi.org/10.1145/3319535.3345663.
Bittel, L. R. (1978). *Encyclopedia of professional management: An authoritative guide to the profitable practice of management.* New York: McGraw-Hill.
Boireau, O. (2018). Securing the blockchain against hackers. *Network Security, 2018*(1), 8–11. https://doi.org/10.1016/S1353-4858(18)30006-0.
Bousquet, A. (2014). Welcome to the machine: Rethinking technology and society through assemblage theory. In S. Curtis (Au.) & M. Acuto (Eds.), *Reassembling international theory: Assemblage thinking and international relations* (pp. 91–97). London: Palgrave Macmillan.
Bui, T., Cooper, D., Collomosse, J., Bell, M., Green, A., Sheridan, J., Higgins, J., Das, A., Keller, J. & Thereaux, O. (2020). Tamper-proofing video with hierarchical attention autoencoder hashing on blockchain. *IEEE Transactions on Multimedia.* https://doi.org/10.1109/TMM.2020.2967640.
Bunge, M. (2012). *Treatise on basic philosophy: Ontology II: A world of systems* (Vol. 4). Dordrecht: Springer.
Burns, C. M., & Hajdukiewicz, J. (2004). *Ecological interface design.* Boca Raton, FL: CRC Press.
Checkland, P. (1999). Systems thinking. In W. L. Currie & B. Galliers (Eds.), *Rethinking management information systems* (pp. 45–56). Oxford: Oxford University Press.
Coase, R. H. (1937). The nature of the firm. *Economica, 4*(16), 386–405. https://doi.org/10.1111/j.1468-0335.1937.tb00002.x.
Cornelius, K. B. (2020). Smart contracts as evidence: Trust, records, and the future of decentralized transactions. In J. Hunsinger, M. M. Allen, & L. Klastrup (Eds.), *Second international handbook of internet research* (pp. 627–646). Dordrecht: Springer. https://doi.org/10.1007/978-94-024-1555-1_28.
De Filippi, P., & Wright, A. (2018). *Blockchain and the law: The rule of code.* Cambridge, MA: Harvard University Press.
Dekker, S. (2016). *Drift into failure: From hunting broken components to understanding complex systems.* Boca Raton, FL: CRC Press.
DuPont, Q. (2019). *Cryptocurrencies and blockchains.* Cambridge: Polity Press.
Eckert, C., & Hillerbrand, R. (2018). Models in engineering design: Generative and epistemic function of product models. In P. E. Vermaas & S. Vial (Eds.), *Advancements in the philosophy of design* (pp. 219–242). Cham: Springer.

Filieri, A., Maggio, M., Angelopoulos, K., d'Ippolito, N., Gerostathopoulos, I., Hempel, A. B., Hoffmann, H., Jamshidi, P., Kalyvianaki, E., Klein, C., Krikava, F., Misailovic, S., Papadopoulos, A. V., Ray, S., Shariffoo, A. M., Shevtsov, S., Ujma, M., & Vogel, T. (2015). Software engineering meets control theory. In *2015 IEEE/ACM 10th International Symposium on Software Engineering for Adaptive and Self-Managing Systems (SEAMS)* (pp. 71–82). Los Alamitos, CA: IEEE Computer Society. https://doi.org/10.1109/SEAMS.2015.12.

Floridi, L. (2005). The ontological interpretation of informational privacy. *Ethics and Information Technology, 7*(4), 185–200. https://doi.org/10.1007/s10676-006-0001-7.

Fox, J. (2007). The uncertain relationship between transparency and accountability. *Development in Practice, 17*(4–5), 663–671. https://doi.org/10.1080/09614520701469955.

Gountia, D. (2019). Towards scalability trade-off and security issues in state-of-the-art blockchain. *EAI Endorsed Transactions on Security and Safety, 5*(18), e4. https://doi.org/10.4108/eai.8-4-2019.157416.

Haller, S., McRoy, S., & Kobsa, A. (Eds.). (2013). *Computational models of mixed-initiative interaction*. Dordrecht: Springer Netherlands. (Original work published 1999)

Harris, D. (Ed.). (2018). *Engineering Psychology and Cognitive Ergonomics: 15th International Conference, EPCE 2018, held as part of HCI International 2018*, Las Vegas, NV, USA, July 15–20, 2018, proceedings (Vol. 10906). Cham: Springer International.

Hirschman, A. O. (1970). *Exit, voice, and loyalty: Responses to decline in firms, organizations, and states*. Cambridge, MA: Harvard University Press.

Hofman, D., Lemieux, V. L., Joo, A., & Batista, D. (2019). The margin between the edge of the world and infinite possibility. *Records Management Journal, 29*(1/2), 240–257. https://doi.org/10.1108/RMJ-12-2018-0045.

Hood, C. (2010). Accountability and transparency: Siamese twins, matching parts, awkward couple? *West European Politics, 33*(5), 989–1009. https://doi.org/10.1080/01402382.2010.486122.

Iguer, H., Medromi, H., Sayouti, A., Elhasnaoui, S., & Faris, S. (2014). The impact of cyber security issues on businesses and governments: A framework for implementing a cyber security plan. In *Proceedings of the 2014 International Conference on Future Internet of Things and Cloud* (pp. 316–321). IEEE. https://doi.org/10.1109/FiCloud.2014.56.

International Organization for Standardization (ISO). (2020, under development). *ISO/FDIS 22739: Blockchain and distributed ledger technologies—vocabulary*.

Kauffman, S. (1995). *At home in the universe: The search for the laws of self-organization and complexity*. Oxford: Oxford University Press.

Kauffman, S. (2013). Evolution beyond Newton, Darwin and entailing law. In B. G. Henning & A. C. Scarfe (Eds.), *Beyond mechanism: Putting life back into biology* (pp. 1–24). Plymouth: Lexington Books.

Kim, C. Y., & Lee, K. (2018). Risk management to cryptocurrency exchange and investors guidelines to prevent potential threats. In *Proceedings of the 2018 International Conference on Platform Technology and Service (PlatCon)* (pp. 1–6). IEEE. https://doi.org/10.1109/PlatCon.2018.8472760.

Korzybski, A. (1933). *Science and sanity: An introduction to non-Aristotelian systems and general semantics*. Lakeville, CT: International Non-Aristotelian Library.

Kroes, P. (2002). Design methodology and the nature of technical artefacts. *Design Studies, 23*(3), 287–302. https://doi.org/10.1016/S0142-694X(01)00039-4.

Lainhart IV, J. W. (2000). COBIT™: A methodology for managing and controlling information and information technology risks and vulnerabilities. *Journal of Information Systems, 14*(s-1), 21–25. https://doi.org/10.2308/jis.2000.14.s-1.21

Lamtzidis, O., & Gialelis, J. (2018). An IOTA based distributed sensor node system. In *Proceedings of the 2018 IEEE Globecom Workshops (GC Wkshps)* (pp. 1–6). IEEE. https://doi.org/10.1109/GLOCOMW.2018.8644153

Latour, B. (1986). Visualisation and cognition: Drawing things together. *Knowledge and Society Studies in the Sociology of Culture Past and Present, 6*, 1–40. Retrieved from http://www.bruno-latour.fr/sites/default/files/21-DRAWING-THINGS-TOGETHER-GB.pdf

Latour, B. (1987). *Science in action: How to follow scientists and engineers through society*. Cambridge, MA: Harvard University Press.

Latour, B. (2005). *Reassembling the social: An introduction to actor-network-theory*. Oxford: Oxford University Press.

Lemieux, V. L. (2016). Trusting records: Is blockchain technology the answer? *Records Management Journal, 26*(2), 110–139. https://doi.org/10.1108/RMJ-12-2015-0042.

Lemieux, V. L. (2017). *Blockchain and distributed ledgers as trusted recordkeeping systems*. Presented at Future Technologies Conference 2017, Vancouver, BC, November 29. Retrieved from https://saiconference.com/Downloads/FTC2017/Proceedings/4_Paper_279-Blockchain_and_Distributed_Ledgers_as_Trusted.pdf

Lemieux, V. L. (2019). Blockchain and public recordkeeping: Of temples, prisons and the (re) configuration of power. *Frontiers in Blockchain, 2*, 1–5. https://doi.org/10.3389/fbloc.2019.00005.

Lemieux, V. L., & Sporny, M. (2017). Preserving the archival bond in distributed ledgers: A data model and syntax. In *WWW '17 Companion: Proceedings of the 26th International Conference on World Wide Web Companion* (pp. 1437–1443). Geneva: International World Wide Web Conferences Steering Committee. https://doi.org/10.1145/3041021.3053896.

Lemieux, V. L., Hofman, D., Batista, D., & Joo, A. (2019). *Blockchain technology & recordkeeping*. ARMA International Education Foundation. Retrieved from http://armaedfoundation.org/wp-content/uploads/2019/06/AIEF-Research-Paper-Blockchain-Technology-Recordkeeping.pdf

Lemieux, V. L., Rowell, C., Seidel M.-D., & Woo, C. (2020). Caught in the middle: Strategic information management disruptions in the era of blockchain and distributed trust. *Records Management Journal*. Advance online publication. https://doi.org/10.1108/RMJ-09-2019-0048.

Lorenz, E. N. (1995). *The essence of chaos*. Seattle: University of Washington Press.

Markey-Towler, B. (2018). *Anarchy, blockchain and utopia: A theory of political-socioeconomic systems organised using blockchain*. Available at SSRN: https://doi.org/10.2139/ssrn.3095343.

May, T. (1992). The crypto anarchist manifesto. In P. Ludlow (Ed.), *High noon on the electronic frontier: Conceptual issues in cyberspace*. Cambridge, MA: MIT Press.

Meyer, M., Sedlmair, M., Quinan, P. S., & Munzner, T. (2015). The nested blocks and guidelines model. *Information Visualization, 14*(3), 234–249. https://doi.org/10.1177/1473871613510429.

Midgley, G. (Ed.). (2003). *Systems thinking*. London: Sage.

Mitzenmacher, M. D. (1996). *The power of two choices in randomized load balancing* (Unpublished doctoral thesis). The University of California, Berkeley, CA. Retrieved from https://www.eecs.harvard.edu/~michaelm/postscripts/mythesis.pdf

Noda, S., Okumura, K., & Hashimoto, Y. (2019). *An economic analysis of difficulty adjustment algorithms in proof-of-work blockchain systems*. Available at SSRN: https://doi.org/10.2139/ssrn.3410460.

Noether, S. (2015). *Ring signature confidential transactions for Monero*. Available from ia.cr/2015/1098

Orlikowski, W. J. (2007). Sociomaterial practices: Exploring technology at work. *Organ Stud, 28* (9), 1435–1448. https://doi.org/10.1177/0170840607081138.

Popper, K. R., & Eccles, J. C. (1977). The worlds 1, 2 and 3. In *The self and its brain: An argument for interactionism* (pp. 36–50). Berlin: Springer.

Rasmussen, J., Pejtersen, A. M., & Goodstein, L. P. (1994). *Cognitive systems engineering*. New York: Wiley.

Rousseau, D. M., Sitkin, S. B., Burt, R. S., & Camerer, C. (1998). Not so different after all: A cross-discipline view of trust. *Academy of Management Review, 23*(3), 393–404.

Searle, J. R. (2006). Social ontology: Some basic principles. *Anthropological Theory, 6*(1), 12–29. https://doi.org/10.1177/1463499606061731.

Senge, P. M. (2006). *The fifth discipline: The art and practice of the learning organization*. London: Random House.

Sheard, S. A., & Mostashari, A. (2009). Principles of complex systems for systems engineering. *Systems Engineering, 12*(4), 295–311. https://doi.org/10.1002/sys.20124.

Skyttner, L. (1996). *General systems theory: An introduction*. Basingstoke: Macmillan Press.

Staples, M. (2014). Critical rationalism and engineering: Ontology. *Synthese, 191*(10), 2255–2279. https://doi.org/10.1007/s11229-014-0396-3.

Staples, M. (2015). Critical rationalism and engineering: Methodology. *Synthese, 192*(1), 337–362. https://doi.org/10.1007/s11229-014-0571-6.

Stewart, J. M., Chapple, M., & Gibson, D. (2012). *CISSP: Certified information systems security professional study guide* (6th ed.). New York: Wiley.

Tejeida-Padilla, R., Badillo-Piña, I., & Morales-Matamoros, O. (2010). A systems science approach to enterprise resources planning systems. *Systems Research and Behavioral Science, 27*(1), 87–95. https://doi.org/10.1002/sres.957.

Vicente, K. J., & Rasmussen, J. (1990). The ecology of human-machine systems II: Mediating 'direct perception' in complex work domains. *Ecological Psychology, 2*(3), 207–249. https://doi.org/10.1207/s15326969eco0203_2.

Vicente, K. J., & Rasmussen, J. (1992). Ecological interface design: Theoretical foundations. *IEEE Transactions on Systems, Man, and Cybernetics, 22*(4), 589–606. https://ieeexplore.ieee.org/document/156574

Vigna, P., & Casey, M. J. (2019). *The truth machine: The blockchain and the future of everything*. New York: Picador.

Von Bertalanffy, L. (1950). An outline of general system theory. *Br J Philos Sci, 1*(2), 134–165. https://doi.org/10.1093/bjps/I.2.134.

Von Bertalanffy, L. (1968). *General system theory: Foundations, development, applications*. New York: Braziller.

Voshmgir, S., & Zargham M. (2019) *Foundations of cryptoeconomic systems*. Cryptoeconomics Working Paper Series: Vienna University of Economics, *1*(1). https://epub.wu.ac.at/7309/

Wade, J., Hoffenson, S., & Gerardo, H. (2019). Systemic design engineering. In E. Bonjour, D. Krob, L. Palladino, & F. Stephan (Eds.), *Complex Systems Design & Management: Proceedings of the Ninth International Conference on Complex Systems Design & Management, CSD&M Paris 2018*. Cham: Springer International.

Walch, A. (2018). Open-source operational risk: Should public blockchains serve as financial market infrastructures? In D. L. K. Chuen & R. Deng (Eds.), *Handbook of blockchain, digital finance, and inclusion, Volume 2: ChinaTech, mobile security, and distributed ledger* (pp. 243–269). New York: Academic. https://doi.org/10.1016/B978-0-12-812282-2.00011-5

Walch, A. (2019a). In code(rs) we trust: Software developers as fiduciaries in public blockchains. In I. Lianos, P. Hacker, G. Dimitripolous, & S. Eich (Eds.), *Regulating blockchain: Techno-social and legal challenges* (pp. 58–81). Oxford: Oxford University Press.

Walch, A. (2019b). Deconstructing 'decentralization': Exploring the core claim of crypto systems. In C. Brummer (Ed.), *Cryptoassets: Legal, regulatory, and monetary perspectives* (pp. 39–68). New York: Oxford University Press.

Williamson, O. E. (1979). Assessing vertical market restrictions: Antitrust ramifications of the transaction cost approach. *University of Pennsylvania Law Review, 127*(4), 953–993. Retrieved from https://scholarship.law.upenn.edu/penn_law_review/vol127/iss4/17

Williamson, O. E. (1986). Vertical integration and related variations on a transaction-cost economics theme. In J. E. Stiglitz & G. F. Mathewson (Eds.), *New developments in the analysis of market structure* (pp. 149–176). London: Palgrave Macmillan.

Woods, D. D., & Roth, E. M. (1988). Cognitive engineering: Human problem solving with tools. *Human Factors, 30*(4), 415–430. https://doi.org/10.1177/001872088803000404.

Yang, S.-B., Lee, K., Lee, H., Chung, N., & Koo, C. (2016). Trust breakthrough in the sharing economy: An empirical study of Airbnb1. In *Proceedings of the 20th Pacific Asia Conference on Information Systems (PACIS 2016)* (p. 131). https://aisel.aisnet.org/pacis2016/131

Yeo, G. (2010). 'Nothing is the same as something else': Significant properties and notions of identity and originality. *Archival Science, 10*(2), 85–116. https://doi.org/10.1007/s10502-010-9119-9.

Zhang, Z., Zargham, M. & Preciado, V. M. (2020). On modeling blockchain-enabled economic networks as stochastic dynamical systems. *Applied Network Science, 5*, art. 19. https://doi.org/10.1007/s41109-020-0254-9

The manufacturer's authorised representative in the EU is Springer Nature Customer Service Centre GmbH, Europaplatz 3, 69115 Heidelberg, Germany. If you have any concerns regarding our products, please contact ProductSafety@springernature.com

Printed and bound by CPI Group (UK) Ltd, Croydon, CR0 4YY
28/09/2026
02224070-0018